Automated Vehicles Symposium 2017

Users. Vehicles. Infrastructure.

San Francisco, California, USA
11-13 July 2017

ISBN: 978-1-5108-5118-4

Printed from e-media with permission by:

Curran Associates, Inc.
57 Morehouse Lane
Red Hook, NY 12571

Some format issues inherent in the e-media version may also appear in this print version.

Printed by Curran Associates, Inc. (2018)

Association for Unmanned Vehicle Systems International
2700 South Quincy Street
Suite 400
Arlington, VA 22206
USA

Phone: (703) 845-9671
Fax: (703) 845-9679

www.auvsi.org

Additional copies of this publication are available from:

Curran Associates, Inc.
57 Morehouse Lane
Red Hook, NY 12571 USA
Phone: 845-758-0400
Fax: 845-758-2633
Email: curran@proceedings.com
Web: www.proceedings.com

TABLE OF CONTENTS

AUTOMATED VEHICLES SYMPOSIUM 2017

USERS. VEHICLES. INFRASTRUCTURE.

Panel Session: Shared Mobility

Moderator:

Kelly Coyner

Senior Fellow, Center for Regional Analysis-Schar School of Policy and Government

George Mason University

Jeff Hobson

Deputy Director for Planning

San Francisco County Transportation Authority

Joseph Okpaku

VP, Government Relations

Lyft

Adam Gromis

Global Lead, Sustainability and Environmental Impact, Policy & Communications

Uber

AUVSI

TRB

Panel Session: Trucking Automation Technology Developments

Moderator: Richard Bishop

Principal

Bishop Consulting

Alden Woodrow

Product Lead, Self Driving Trucks

Uber

Michael Cammisa

VP, Safety & Policy & Connectivity

American Trucking Associations

Dr. Kelly Regal

Associate Administrator of Research and Information Technology

FMCSA

Max Fuller

Executive Chairman

U.S. Xpress, Inc.

Dr. Aravind Kailas

Principal Technology Planner

Volvo Group North America

Dr. Josh Switkes

Founder & CEO

Peloton Technology

Caltrans Leadership in Automated Vehicle Research

Malcolm Dougherty
Director
California Department of Transportation
(Caltrans)

July 11, 2017

Additional Transportation Funding (Senate Bill 1)

Provides more than $50 billion for improvements to California's transportation infrastructure over the next ten years, including increased funding for the State Highway System:

- $15 billion for state highway maintenance and rehabilitation
- $4 billion for bridge and culvert maintenance and rehabilitation
- $3 billion for high priority freight corridors
- $2.5 billion for congested corridor relief

DMV Regulations for Testing and Operation of AV's

Testing:

- Became effective on September 16, 2014
- 36 companies are currently permitted to test

Operation: (with some changes to testing also)

- Formal proposed rules published on March 10, 2017
    - The rule making process is progressing
    - "Driverless" testing and operation would be allowed

Please see:

www.dmv.ca.gov/portal/dmv/detail/vr/autonomous/auto

Demo '97

ACC versus Cooperative ACC

Truck Automation Testing

Partial Automation for Truck Platooning

Demo 2003 (Automated Bus Platooning)

Vehicle Assist & Automation (Bus Steering Automation)

Snow Plow Operator Assistance

Thanks for Participating

Caltrans provides a safe, sustainable, integrated and efficient transportation system to enhance California's economy and livability

NISSAN
Innovation that excites

Integrating Autonomous Drive into the New Automotive Reality

Maarten Sierhuis, Ph.D.
Nissan Research Center Silicon Valley

www.nissan-global.com

Nissan's Autonomous drive technology development

NISSAN

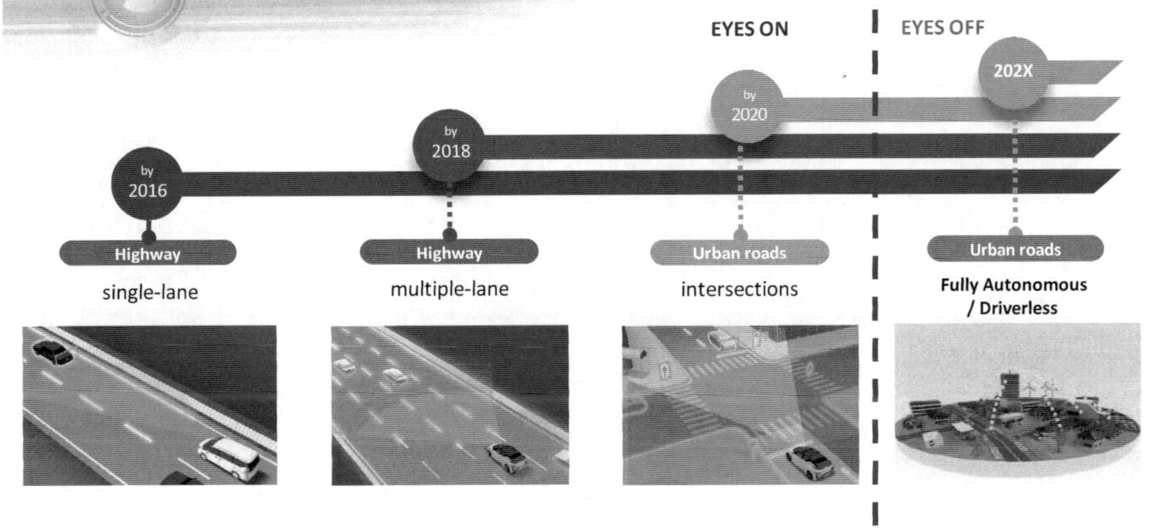

NISSAN

Socially Acceptable Autonomy

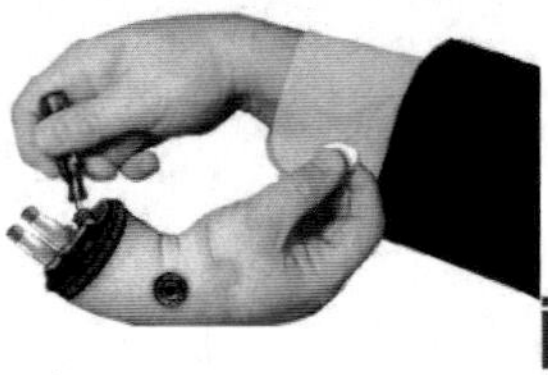

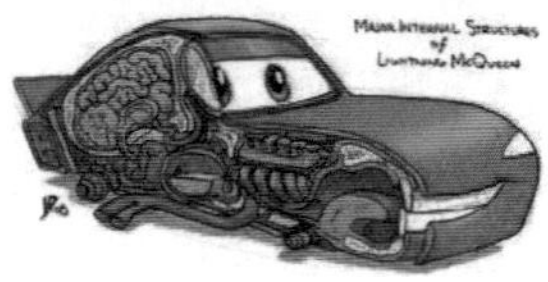

- **Social** → Humans are SOCIAL
- **Interaction** → People INTERACT and COMMUNICATE
- **Inside the car**
  - Social → What services does the car provide to the human and vice-versa?
  - Interaction → People inside and the car interact
- **Outside the car**
  - Social → The car needs to understands other road users
  - Interaction → People outside and the car interact

NISSAN

Coping with real-life mobility situations

NISSAN

ARTCC - Air Route Traffic Control Centers

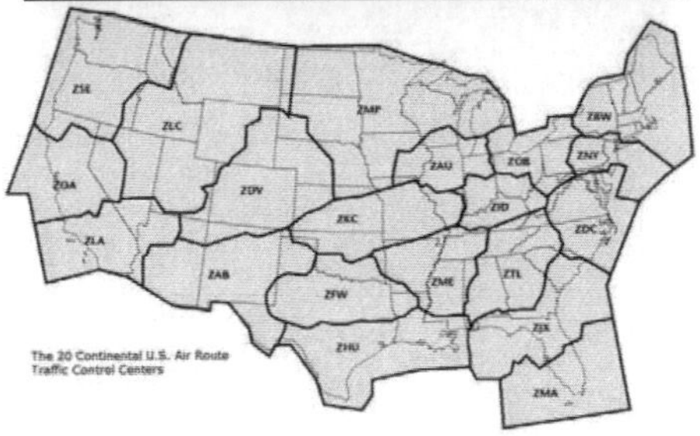

The 20 Continental U.S. Air Route Traffic Control Centers

Air traffic controllers are personnel responsible for the **safe, orderly, and expeditious flow** of air traffic in the global air traffic control system.

Usually stationed in **air traffic control centers** and control towers on the ground, they **monitor the position, speed, and altitude** of aircraft in their assigned airspace visually and by radar, and **give directions to the pilots by radio**.

The position of air traffic controller is one that requires **highly specialized knowledge, skills, and abilities**.

NISSAN

SAM - Seamless Autonomous Mobility

Developed from NASA Technology

- Edge cases that need human perception, understanding and decision.
- No fully autonomous system works without **human-in-the-loop**
- **Human-robot teaming** brings the best of the two intelligences
- **Human safety net** is expected by users

→ **SAM concept**

- A human in the loop enhances the autonomous capabilities of an AV Fleet
- AI in the cloud assists operators and Avs to perform more efficiently

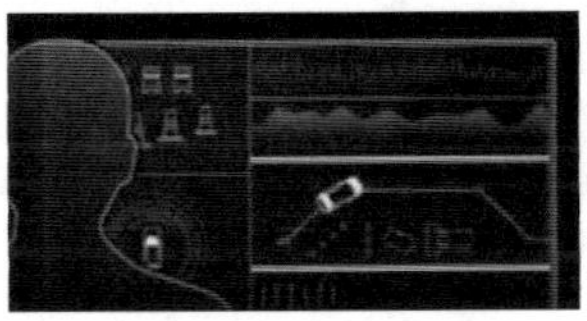

Tele Op

+

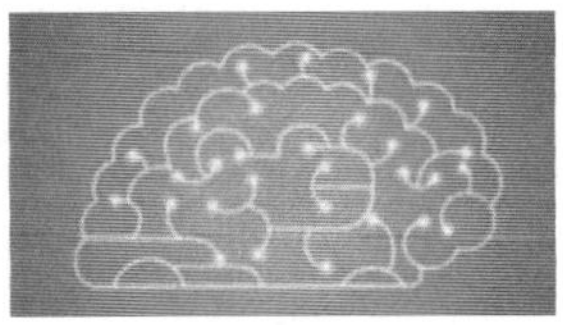

A.I. in the Cloud

NISSAN

Mobility Control Center

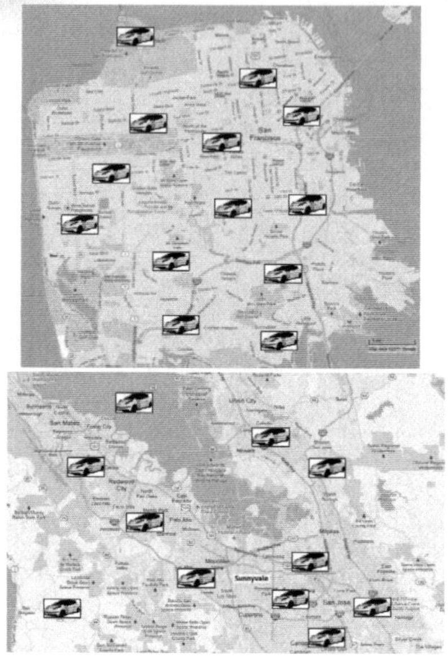

Mobility Managers are personnel responsible for the **safe, orderly, and expeditious flow** of **Autonomous Fleets** in the global traffic system.

NISSAN

Distributed Artificial Intelligence with Human-In-The-Loop

Individual Autonomous Systems

Human – Autonomous System

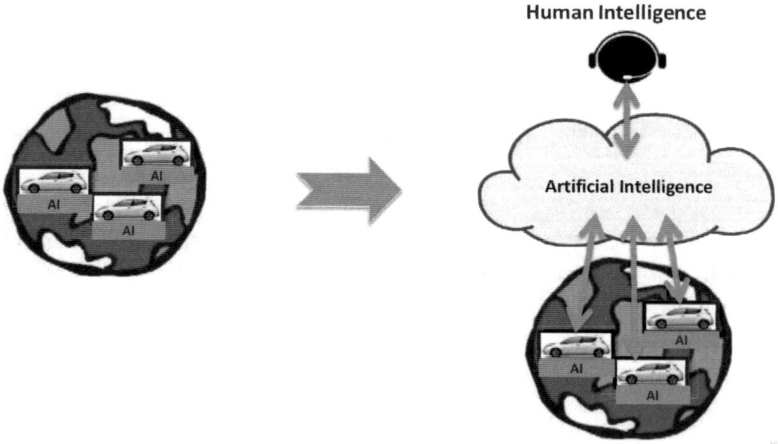

SAM

Developed from NASA Technology

NISSAN

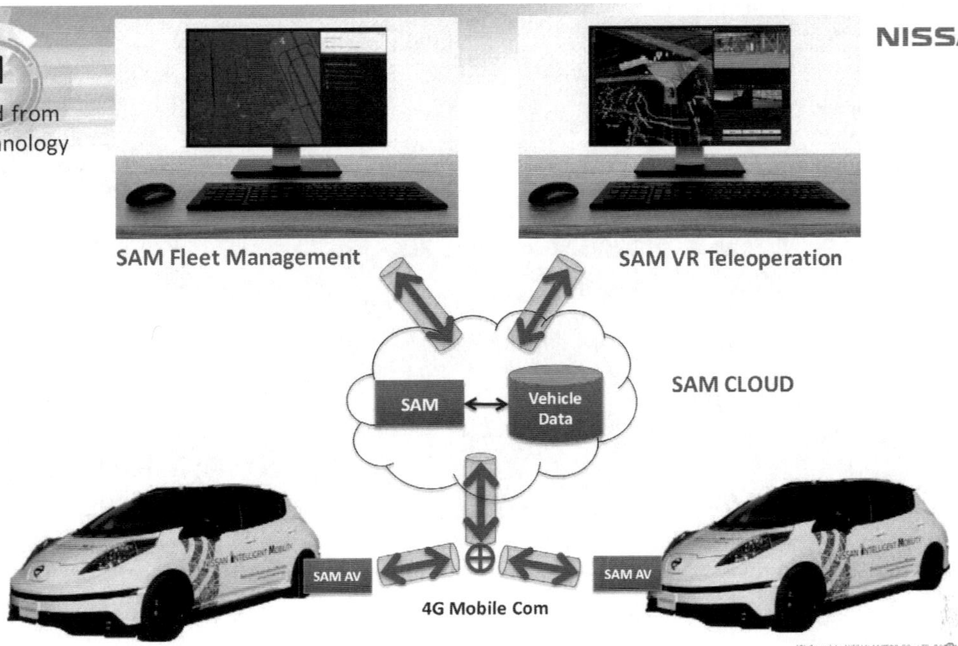

SAM Functions

Developed from NASA Technology

NISSAN

SAM Fleet Supervision
- Fleet status
- Tele-operation zones
- Vehicles requesting help
- Intervention history

SAM Vehicle Supervision
- Situational awareness
  - 3D Virtual view
  - Camera (zoom and frame-rate)
  - Lidar
- "Paint" a Solution

SAM AV
- Vehicle to cloud telemetry
- Tele-op request
- Solution execution

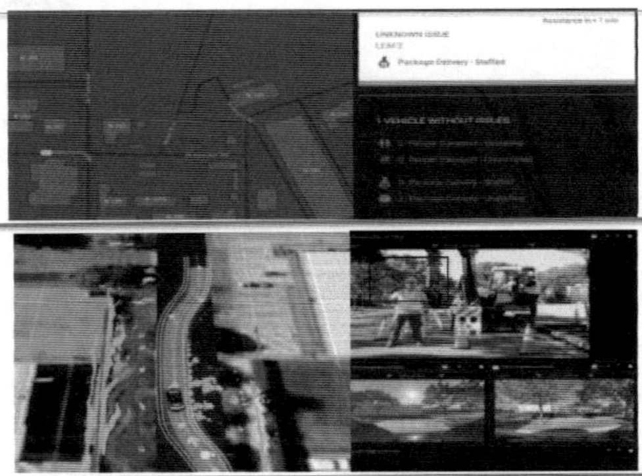

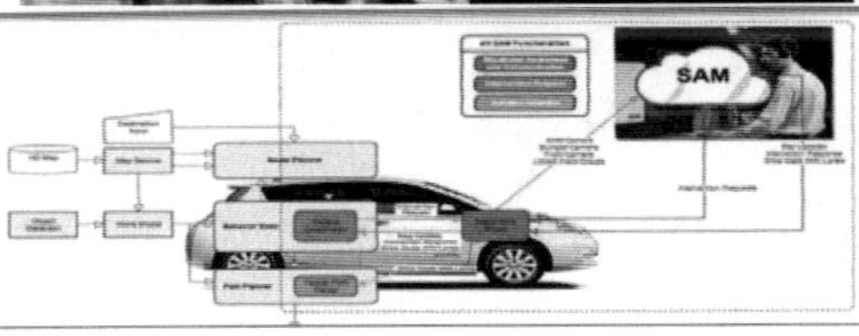

Communication

NISSAN

SAM – The Mobility Manager

Developed from NASA Technology

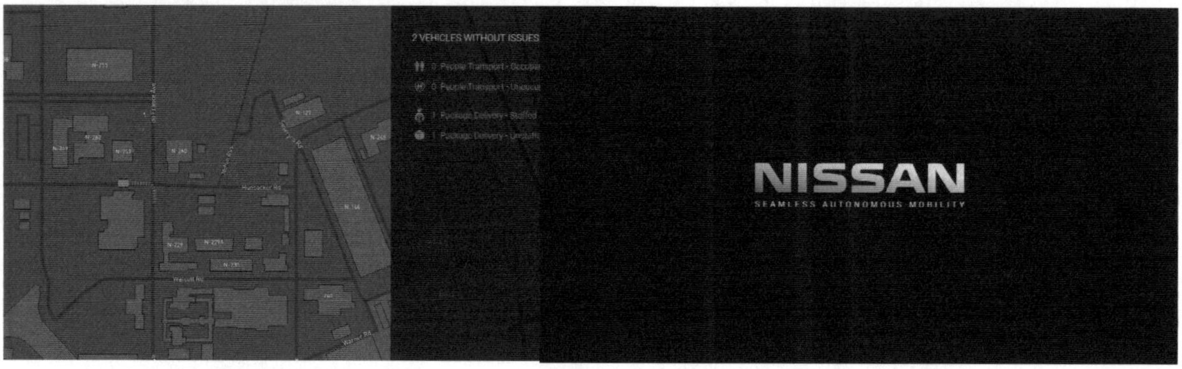

11

NISSAN

Thank you for your attention

RAND
CORPORATION
Regulating Autonomous Vehicles
Amid Uncertainty
Nidhi Kalra
Image by H. Miller.

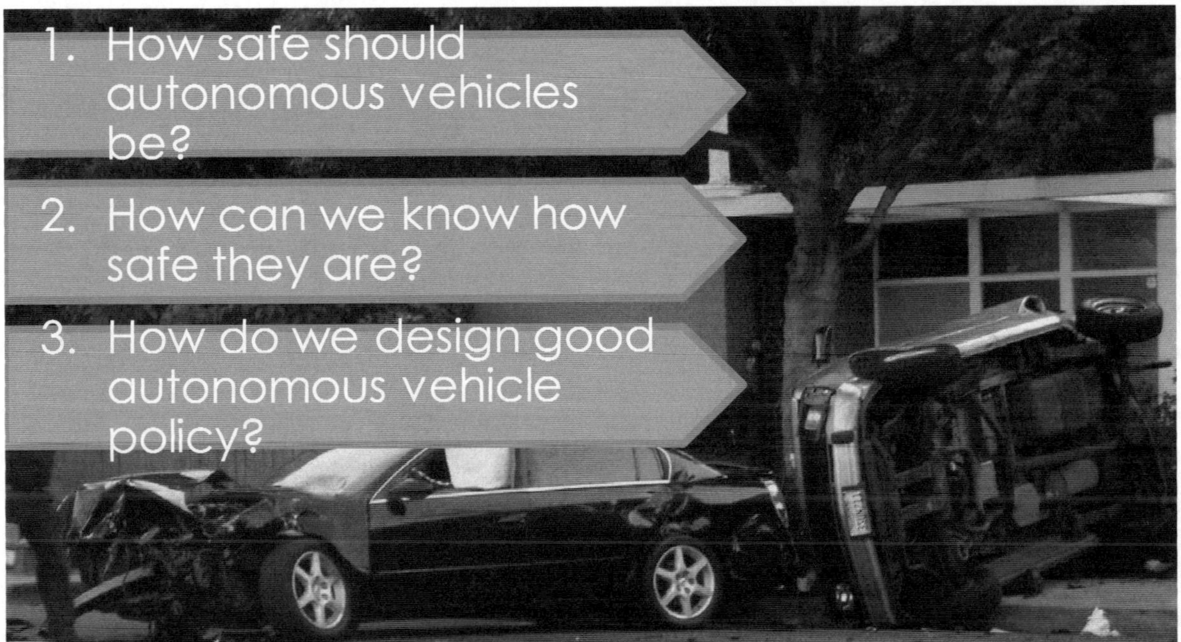
1. How safe should autonomous vehicles be?
2. How can we know how safe they are?
3. How do we design good autonomous vehicle policy?

1. How safe should autonomous vehicles be?
Uncertain
2. How can we know how safe they are?
3. How do we design good autonomous vehicle policy?

1. How safe should autonomous vehicles be?
Uncertain
2. How can we know how safe they are?
Uncertain
3. How do we design good autonomous vehicle policy?

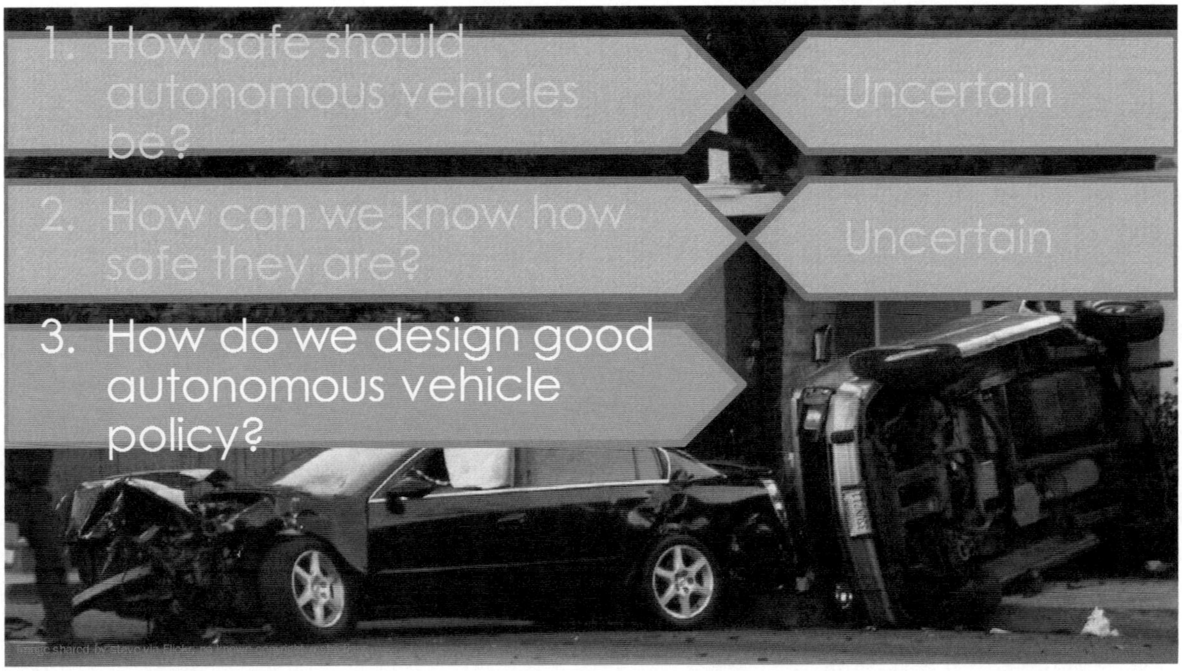

Federal Motor Vehicle Safety Standards (FMVSS) pose a barrier to deploying some types of AVs

- FMVSS specify car design, construction, performance, and durability requirements
- Many innovative AV designs (e.g., no human driver) would not comply with FMVSS

NHTSA offers exemptions to FMVSS for development of innovative safety features

But NHTSA limits exposure to risk by requiring certain conditions.

Safety

Developers must demonstrate equivalent safety between conforming and non-conforming vehicles

Credit: Brady Holt

Quantity

Developers may obtain exemption for only 2,500 vehicles per year

Credit: automobileitalia

"Practical Autonomous Vehicle Exemptions" or "PAVE Act" proposes raising FMVSS exemptions from 2,500 to **100,000 vehicles/exemption/year.**

Does this make sense?

Neither the letter nor the spirit of the exemption process works when applied to AVs

Need to demonstrate that the non-conforming vehicle can be as safely driven by a human as a conforming vehicle

Neither the letter nor the spirit of the exemption process works when applied to AVs

Need to demonstrate that the non-conforming vehicle can be as safely driven by a human as a conforming vehicle

Need to demonstrate that an autonomous vehicle can drive itself safely

Neither the letter nor the spirit of the exemption process works when applied to AVs

Need to demonstrate that the non-conforming vehicle can be as safely driven by a human as a conforming vehicle

Need to demonstrate that an autonomous vehicle can drive itself safely

- No definition of AV safety

Neither the letter nor the spirit of the exemption process works when applied to AVs

Need to demonstrate that the non-conforming vehicle can be as safely driven by a human as a conforming vehicle

Need to demonstrate that an autonomous vehicle can drive itself safely

- No definition of AV safety
- No practical way to demonstrate safety prior to deployment

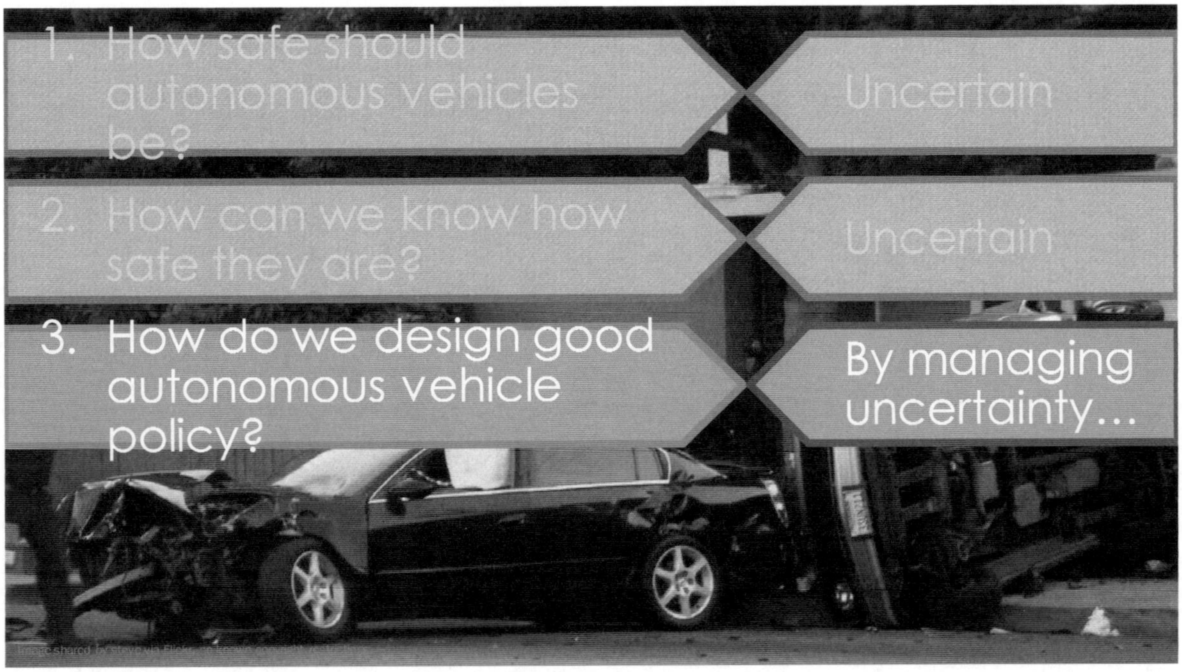

There is a tradeoff between risk and reducing uncertainty

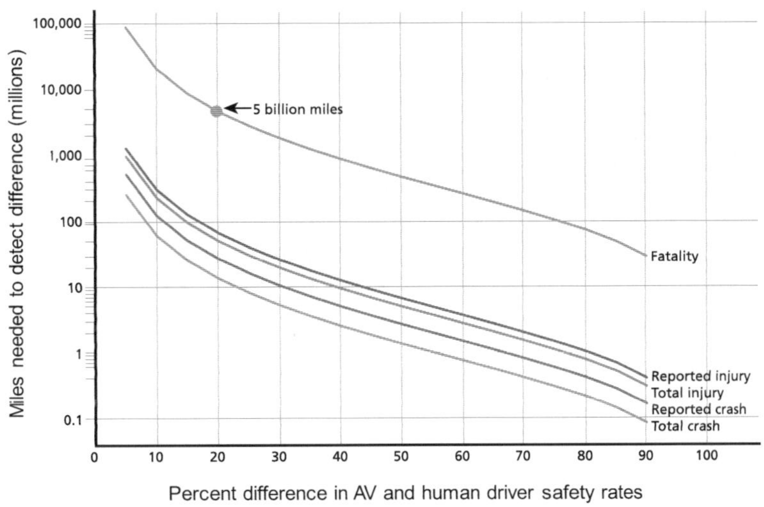

Kalra, Nidhi and Susan Paddock. Driving to Safety: How Many Miles of Driving Would It Take to Demonstrate Autonomous Vehicle Reliability?. Santa Monica, CA: RAND Corporation, 2016. https://www.rand.org/pubs/research_reports/RR1478.html.

What if regulations were designed with this risk-information tradeoff in mind?

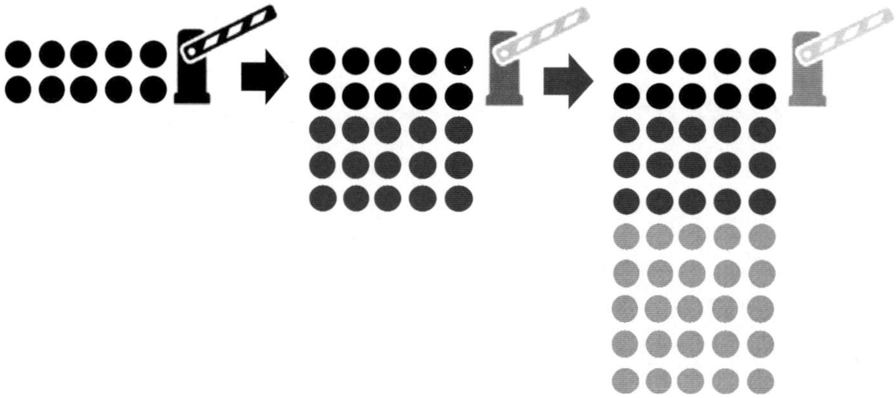

This tradeoff can be understood mathematically

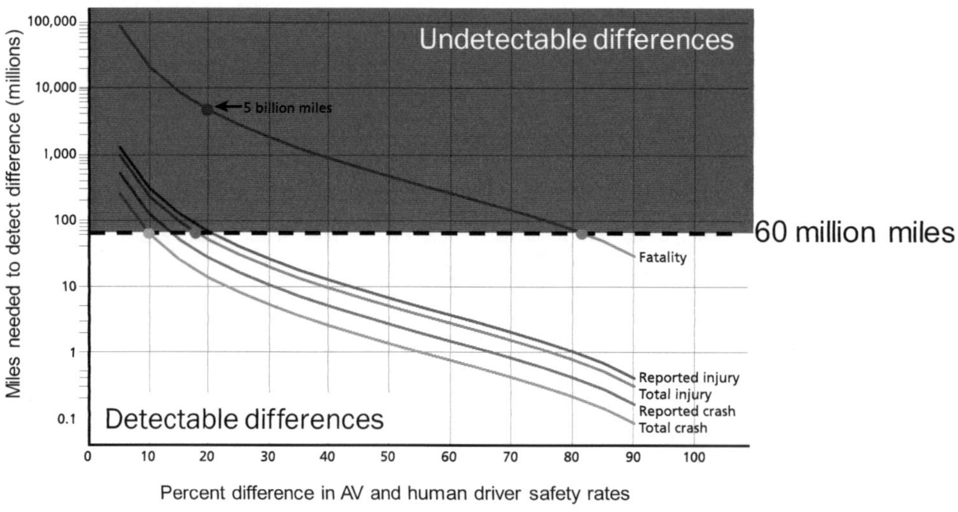

This tradeoff can be understood mathematically

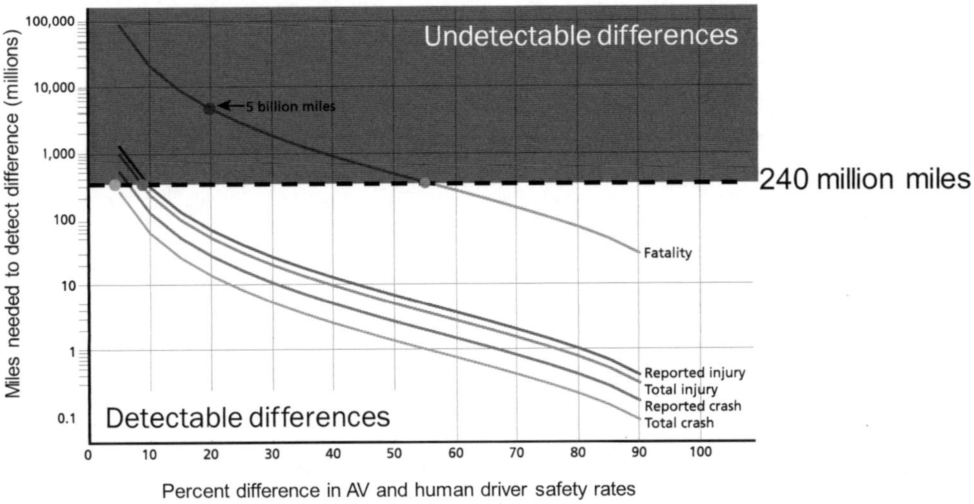

Bottom line

- When faced with great uncertainty, it is almost impossible to get regulations right the first time
- Policies that manage uncertainty could enable innovation, while balancing the tradeoff between risk and information
- A graduated approach additionally helps avoid the problem of the horse and the barn door

RAND
CORPORATION

Ministry of Infrastructure and the Environment

Vehicle Regulation in Europe

Global, EU and national legislation for Connected and Automated Driving

Drs. Edwin Nas

Dep. Projectleader Connected and Automated Driving The Netherlands

Rapporteur for the EU on CAD (GEAR2030)

17 July 2017

Make the EU type approval system future proof

Declaration of Amsterdam
Cooperation in the field of connected and automated driving
14-15 April 2016
Informal Transport Council
THE NETHERLANDS
EU
2016
3
Ministry of Infrastructure and the Environment
17 July 2017

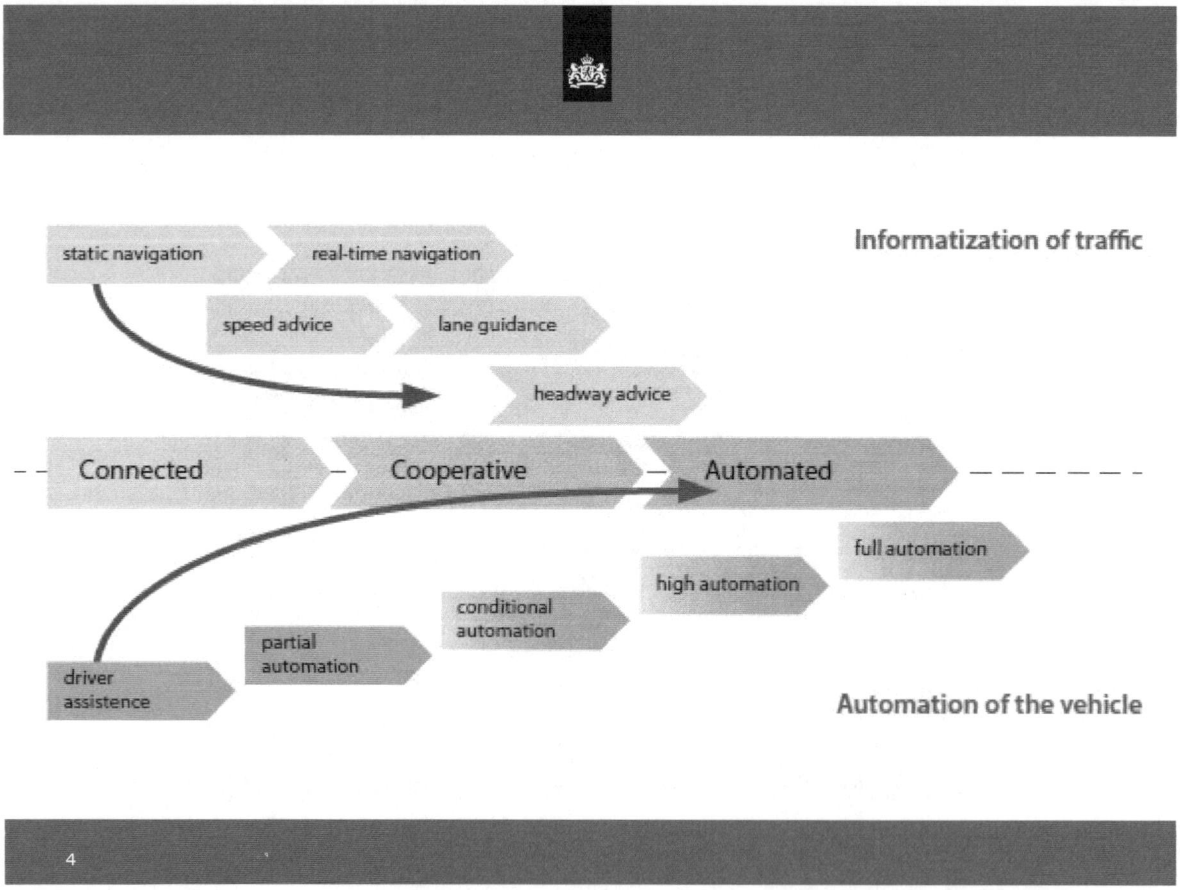
static navigation
real-time navigation
speed advice
lane guidance
headway advice
Informatization of traffic
Connected
Cooperative
Automated
full automation
high automation
conditional automation
partial automation
driver assistence
Automation of the vehicle
4

Structural high level meetings EU

Ministry of Infrastructure and the Environment

On our way towards connected and automated driving in Europe

Outcome of the first High Level Meeting

Amsterdam, 15 February 2017

Analysis

- Vehicle becomes 'ever changing'
- Use phase is becoming more important because here is where the road safety is determined: focus on surveillance
- Software evolves iteratively (and OTA)

- Approach: Performance Based Standards & Acceptable Means of Compliance

So: regulations should describe the 'what', not the 'how'
Innovations should be proven safe and can be added to AMC

New approach

Transition schemes

1. Data: from static – to dynamic (place and time-bound)
2. Sharing and learning from accidents and incidents
3. Basis of data: from a central database – to the vehicle
4. Driving: from human – to software
5. Approval: from admittance – to admittance with selfassessment on innovative systems
6. Surveillance: from periodic – to continuous/data driven
7. Road safety: from in the vehicle – to the interaction between vehicles and infrastructure
8. Responsibility: from driver to vehicle – Driver Level of Responsibility (DLR)

Two-phased approach to legislating

Phase 1: creating the possibilities for development

- Opening the door for CAD / AV
- Defining legal barriers for market introduction
- Address the societal implications

Phase 2: legislating the right requirements

- Using experience of testing and experiments
- Knowing what to regulate
- Optimizing choice for functional of technical requirements and standards

9

Global, EU and national framework

UN-ECE wants to move forward on CAD

- Role of the driver is related to new vehicle systems

When the vehicle is driven by vehicle systems that do not require the driver to perform the driving task, the driver can engage in ***activities other than driving*** *as long as:*

- *Principle 1: these activities do not prevent the driver from responding to demands from the vehicle systems for taking over the driving task, and*
- *Principle 2: these activities are consistent with the prescribed use of the vehicle systems and their defined functions.*

- New guidance document for highly automated/driverless vehicles addressing both Geneva and Vienna Conventions

EU program: GEAR2030

High Level Group GEAR 2030

for 2 years '16-'17, composed by industry, NGOs and Member States.

Objective: build a coherent approach on the industrial development of connected and automated vehicles.

- Working group 1: Adaptation of the EU Value Chain
- Working group 2: *Connected and Automated Driving*
- Working group 3: Global Competitiveness

Strategy 2030 Report expected October 2017

WG2: connected and automated driving (CAD)

- Dedicated working group with car manufacturers, suppliers, member states, insurance industry, telecom providers, NGO's

- Focus on policy, regulatory and financing issues.
- First recommendations for upcoming systems (2020) ready.
- Final long-term (2030) recommendations by September 2017.

Towards 2030

- The EU framework already provides a relevant framework for automated and connected cars expected for 2020.

- For 2030 we will also elaborate on:
- New vehicle approval system in the EU, including innovative systems
- New liability scemes for higher levels of autonomy
- Data in vehicles and data-access related to privacy
- Working with voluntary building blocks for testing, towards mutual recognition
- Helping countries with legislation

Upcoming systems (2020)

- Upcoming systems are: mass market 2020 systems at SAE level 2-3-4 such as:
- - motorway (high speed): highway pilot, platooning
- - city (lower speed): traffic jam assist, manoeuvres at low speed, trips in some dedicated/secured areas
- - dedicated POD-vehicles in separate areas and on open road (level 4/5)
- In the perspective of international and cross border traffic

Large scale Cross Border testing

16

National approach

- Open road testing with driver possible nationwide (2015)
  - One stop shop for exemption at rdw.nl/its
  - Driver in or outside the vehicle is within the law
  - No legal reference to holding the steering wheel
  - Distance keeping is functional: "safe distance"

- New experimental law nationwide (per 2018)
  - Driving on open road without a driver inside the vehicle
  - New approach for the term 'driver'
  - New testing possibilities for highly and fully autonomous

17 Ministry of Infrastructure and the Environment 17 July 2017

18 Ministry of Infrastructure and the Environment 17 July 2017

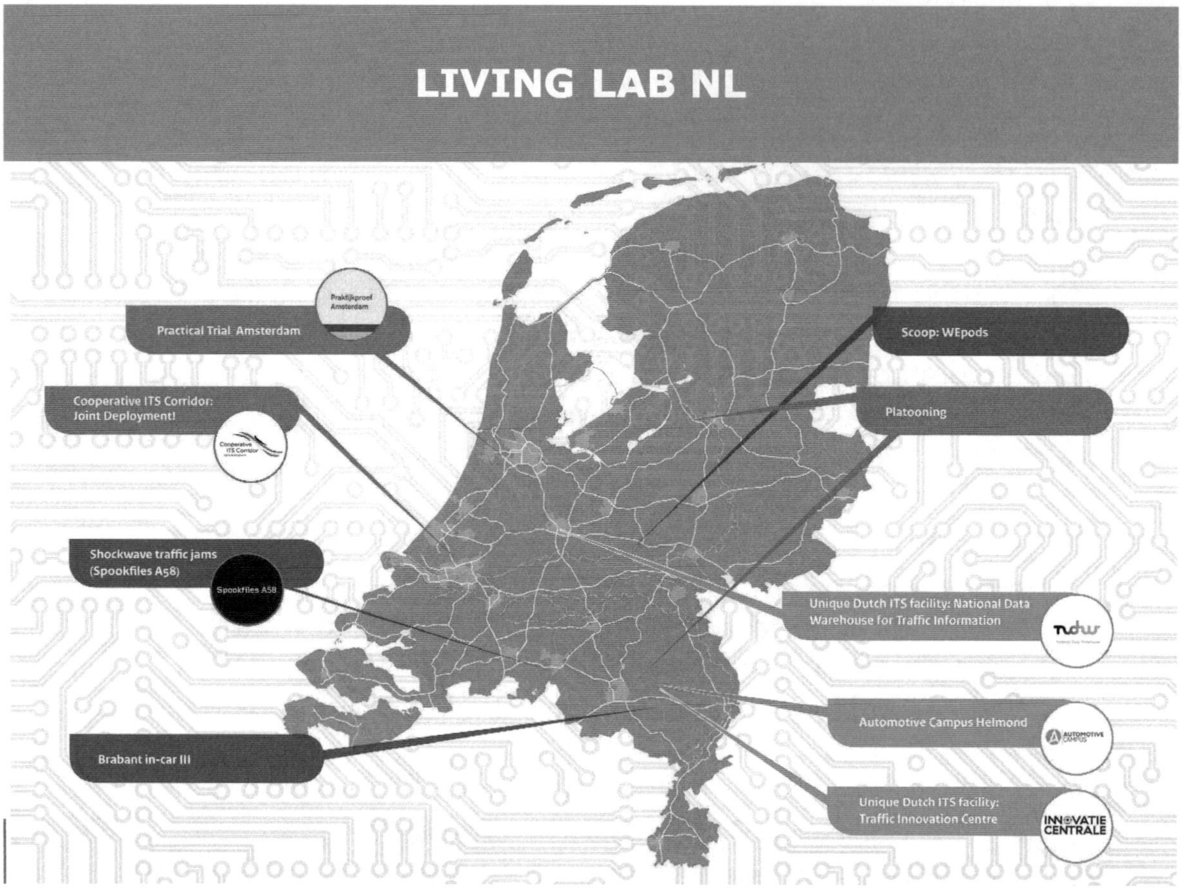
LIVING LAB NL
Practical Trial Amsterdam
Praktijkproef Amsterdam
Cooperative ITS Corridor: Joint Deployment!
Cooperative ITS Corridor
Shockwave traffic jams (Spookfiles A58)
Spookfiles A58
Brabant in-car III
Scoop: WEpods
Platooning
Unique Dutch ITS facility: National Data Warehouse for Traffic Information
Automotive Campus Helmond
Unique Dutch ITS facility: Traffic Innovation Centre
INNOVATIE CENTRALE

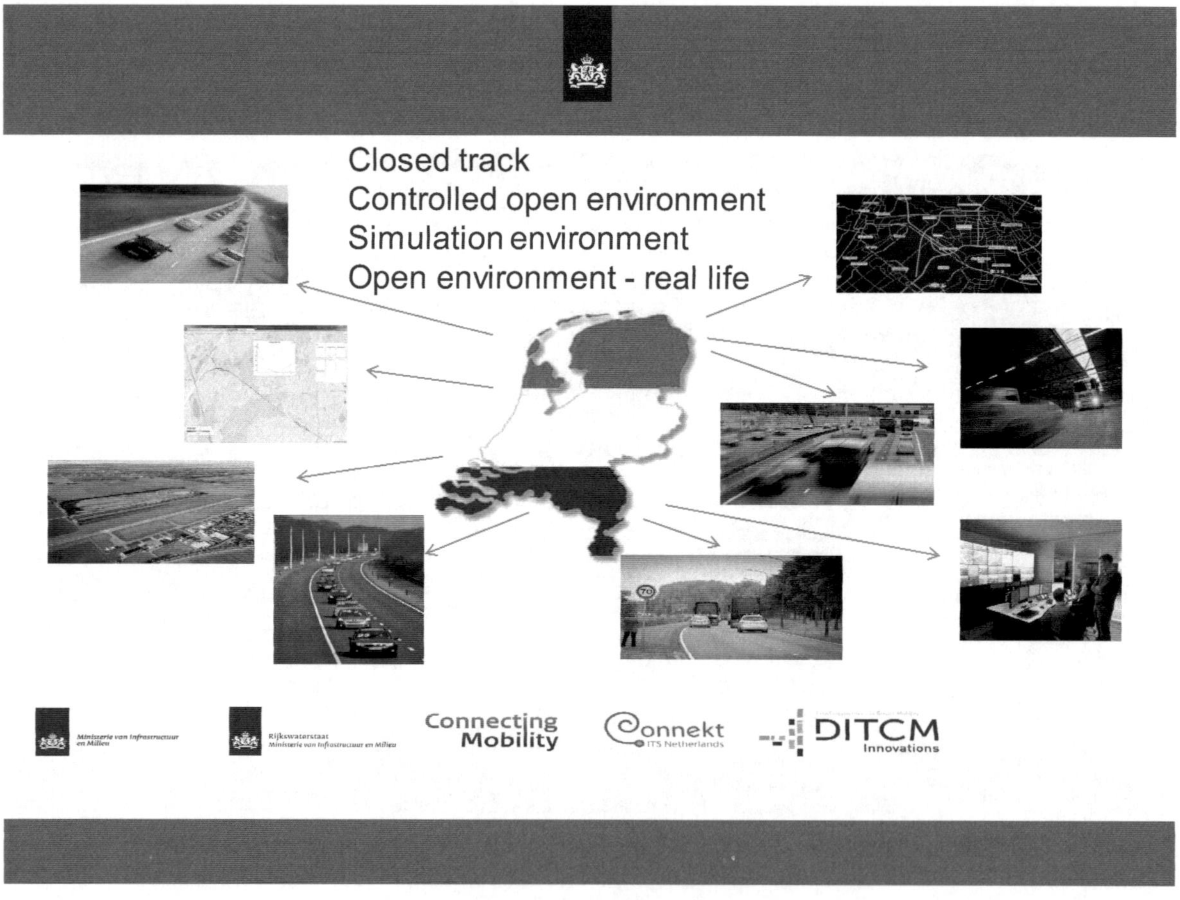
Closed track
Controlled open environment
Simulation environment
Open environment - real life
Ministerie van Infrastructuur en Milieu
Rijkswaterstaat
Ministerie van Infrastructuur en Milieu
Connecting Mobility
Connekt
ITS Netherlands
DITCM
Innovations

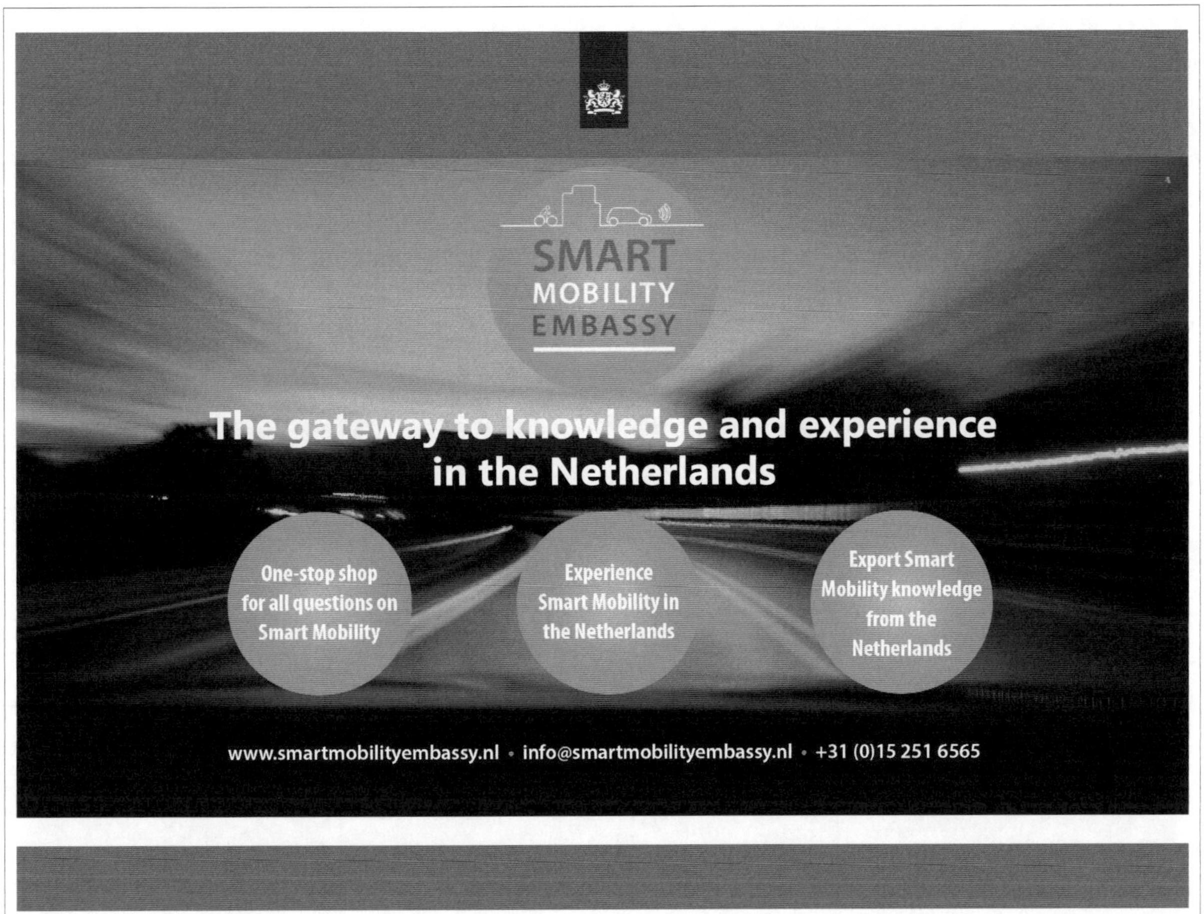
SMART
MOBILITY
EMBASSY
The gateway to knowledge and experience in the Netherlands
One-stop shop for all questions on Smart Mobility
Experience Smart Mobility in the Netherlands
Export Smart Mobility knowledge from the Netherlands
www.smartmobilityembassy.nl • info@smartmobilityembassy.nl • +31 (0)15 251 6565

USE OUR FACILITIES FOR COOPERATIVE, CONNECTED AND AUTOMATED DRIVING IN ENVIRONMENTS LIKE:
CLOSED TRACK
CONTROLLED
SIMULATION
OPEN - REAL LIFE
...AND MORE
Why the Netherlands?
• The Netherlands offers one of the best infrastructures in the world - it tops the rankings in terms of connectivity, offers ITS using high-speed broadband and has 98% household coverage and national 4G coverage
• Access to the complete testing chain
• Strong public-private cooperation
• Gateway to Europe for legislative changes
• Engaged customers: 80% of Dutch people have a smartphone and the logistics sector is involved in the platooning challenge.

Sharing=caring

KENNISAGENDA

Search ...

KENNISDOMEINEN BIBLIOTHEEK EVENEMENTEN PROEVEN FILMPJES FAQ

ENGLISH

Click here for the English version of the knowledge agenda

CONTACT

Als u een bijdrage wilt leveren (rapporten, presentaties, congressen, etc.) dan kunt u dat doen via een mailtje naar: tom.alkim@rws.nl . Voor vragen kunt u ook kijken op de LinkedIn pagina.

Welkom bij Kennisagenda Automatisch Rijden, een initiatief van de werkgroep de ZelfRijdende Auto (ZRA), van het Ministerie van Infrastructuur en Milieu, Rijkswaterstaat en de RDW, om een online overzicht te geven van beschikbare en benodigde kennis op het gebied van automatisch rijden. Het overzicht is verdeeld in een aantal kennisdomeinen om de diverse facetten in beeld te brengen. In de bibliotheek vindt u een uitgebreide collectie van rapporten, papers en presentaties, inclusief samenvattingen en achtergrondinformatie.
Op 14 april 2015 is de kennisagenda van de werkgroep ZRA gepresenteerd en is in een aantal workshops de kennisbehoefte per domein in kaart gebracht. Op deze site vindt u ook een overzicht van relevante congressen en evenementen en een collectie filmpjes en webinars. Nieuws en actuele ontwikkelingen worden middels de bibliotheek en twitterfeed (#KARNL) door ons bijgehouden.

http://knowledgeagenda.connekt.nl

23

Let's put the pieces together...

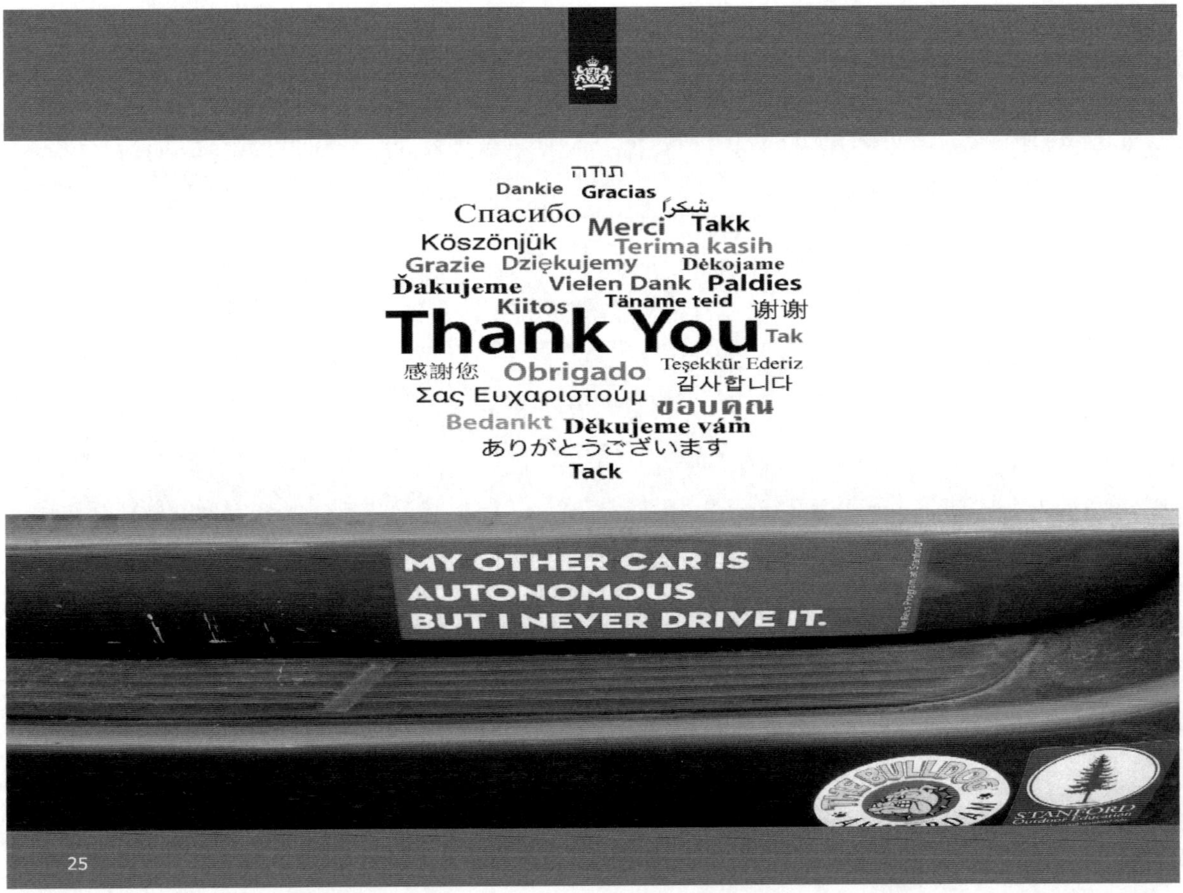
תודה
Dankie
Gracias
Спасибо
شكرا
Merci
Takk
Köszönjük
Terima kasih
Grazie
Dziękujemy
Dėkojame
Ďakujeme
Vielen Dank
Paldies
Kiitos
Täname teid
谢谢
Thank You
Tak
感謝您
Obrigado
Teşekkür Ederiz
감사합니다
Σας Ευχαριστούμ
ขอบคุณ
Bedankt
Děkujeme vám
ありがとうございます
Tack
MY OTHER CAR IS
AUTONOMOUS
BUT I NEVER DRIVE IT.
25

More info:
Edwin Nas – edwin.nas@minienm.nl
+31 6 15 35 94 11

Deep Learning & Autonomous Vehicles

motus ventures

Robert Seidl
robert@motusventures.com

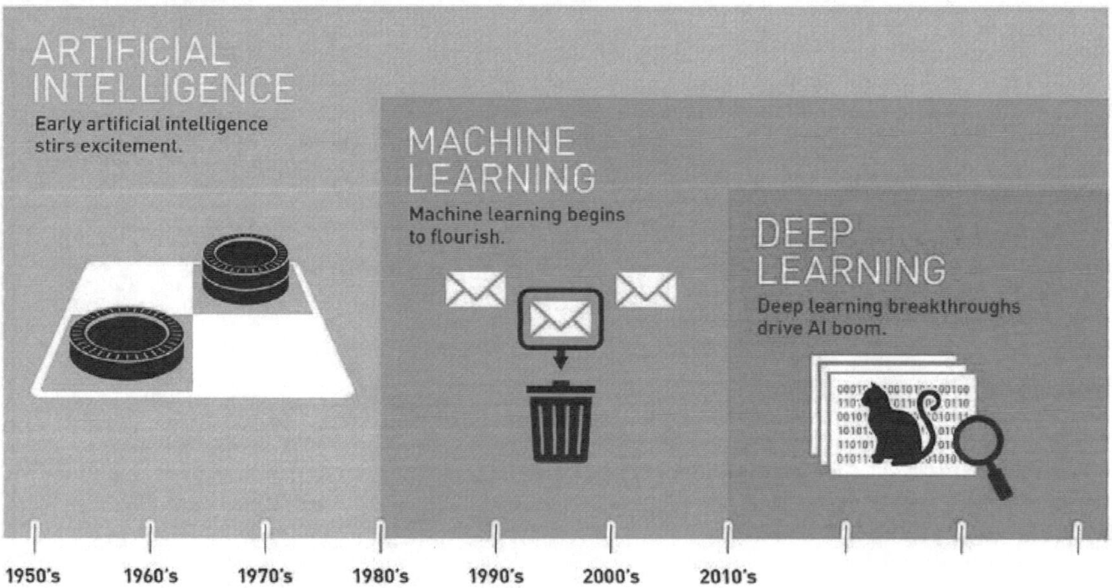

Since an early flush of optimism in the 1950s, smaller subsets of artificial intelligence – first machine learning, then deep learning, a subset of machine learning – have created ever larger disruptions.

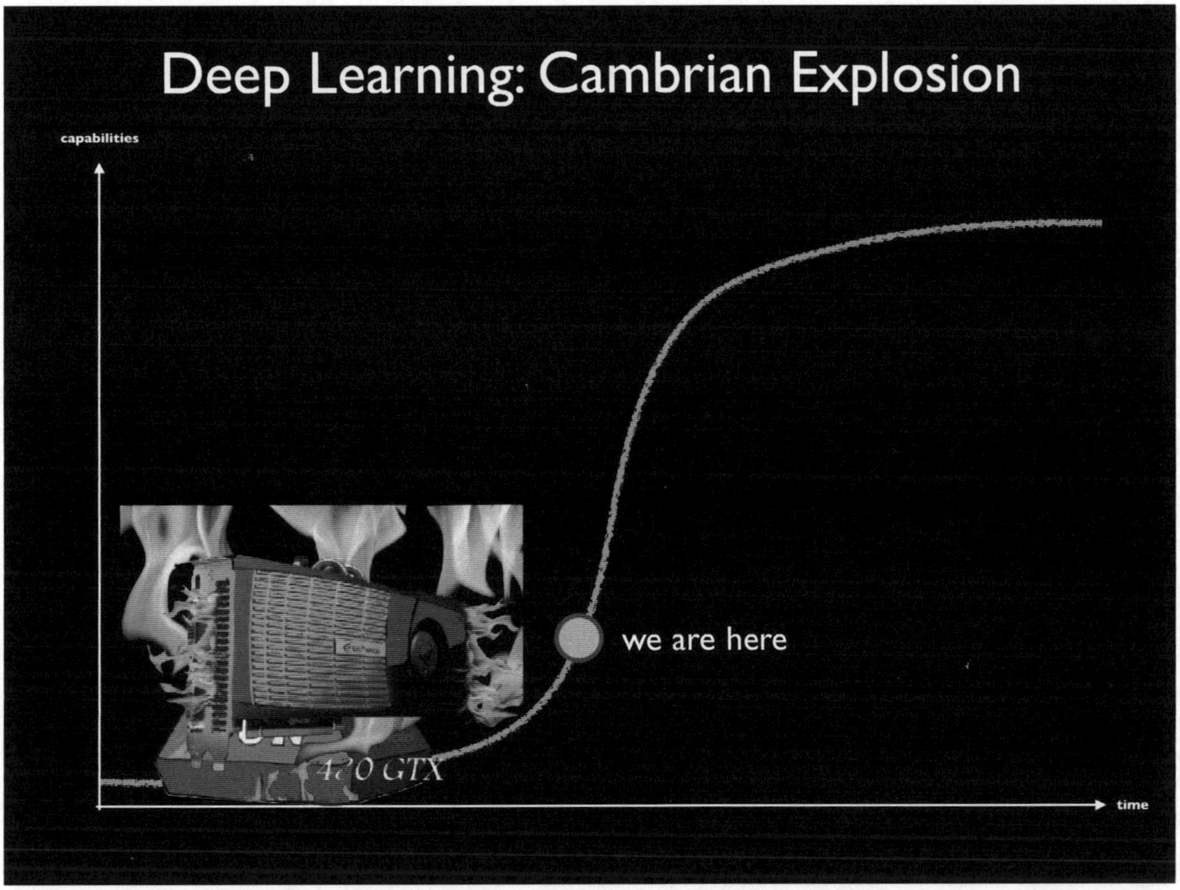
Deep Learning: Cambrian Explosion
capabilities
we are here
420 GTX
time

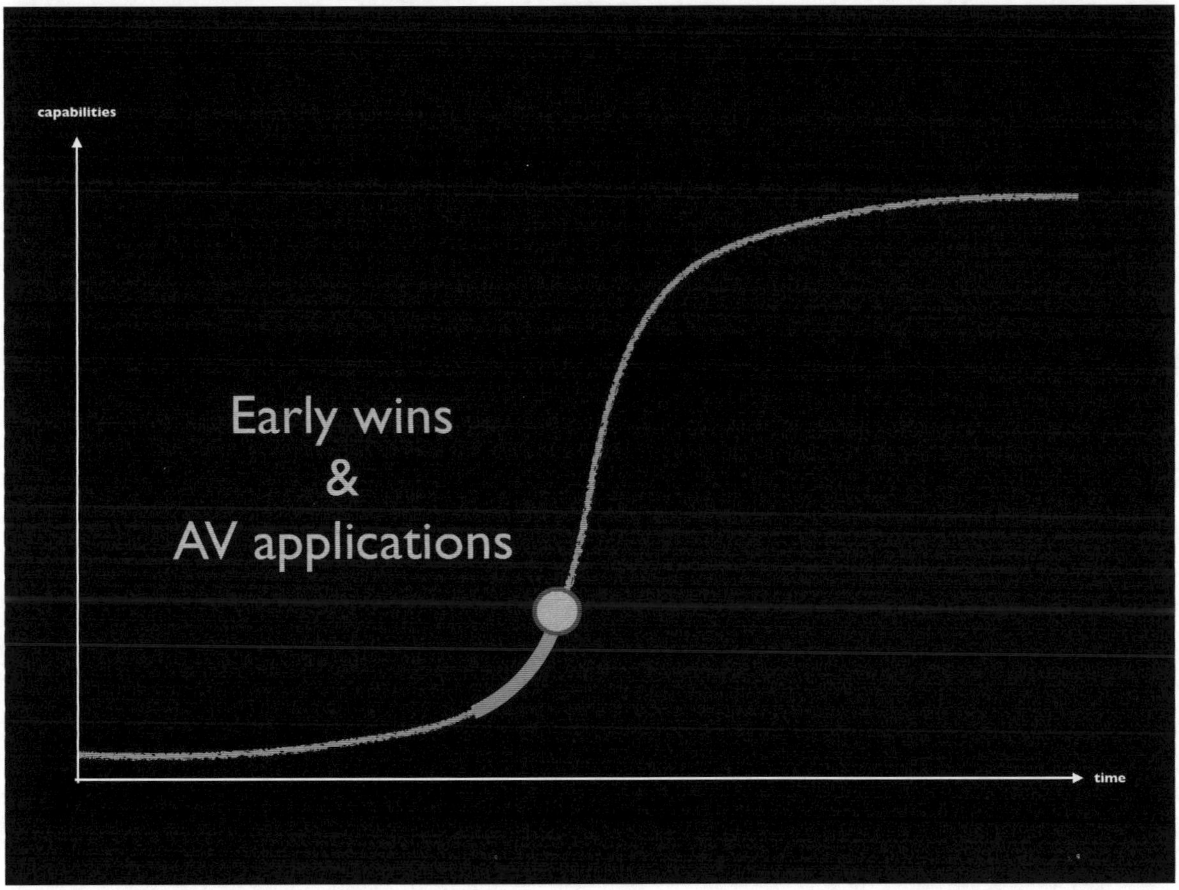
capabilities
Early wins
&
AV applications
time

Deep Learning Success Stories

- Object recognition (cats, faces, cars, voices etc.)
- Image matching, Machine translation, Games, Lip reading
- Healthcare: iphone dermatology diagnostics, Deep Patient
- Generative networks
  - Voice synthesis
  - Chat bots
  - Image Captioning
  - Image completion, inpainting

Object detection & bounds

Road & Lane markings

Image Segmentation

Free space detection
Car
Car side
Road

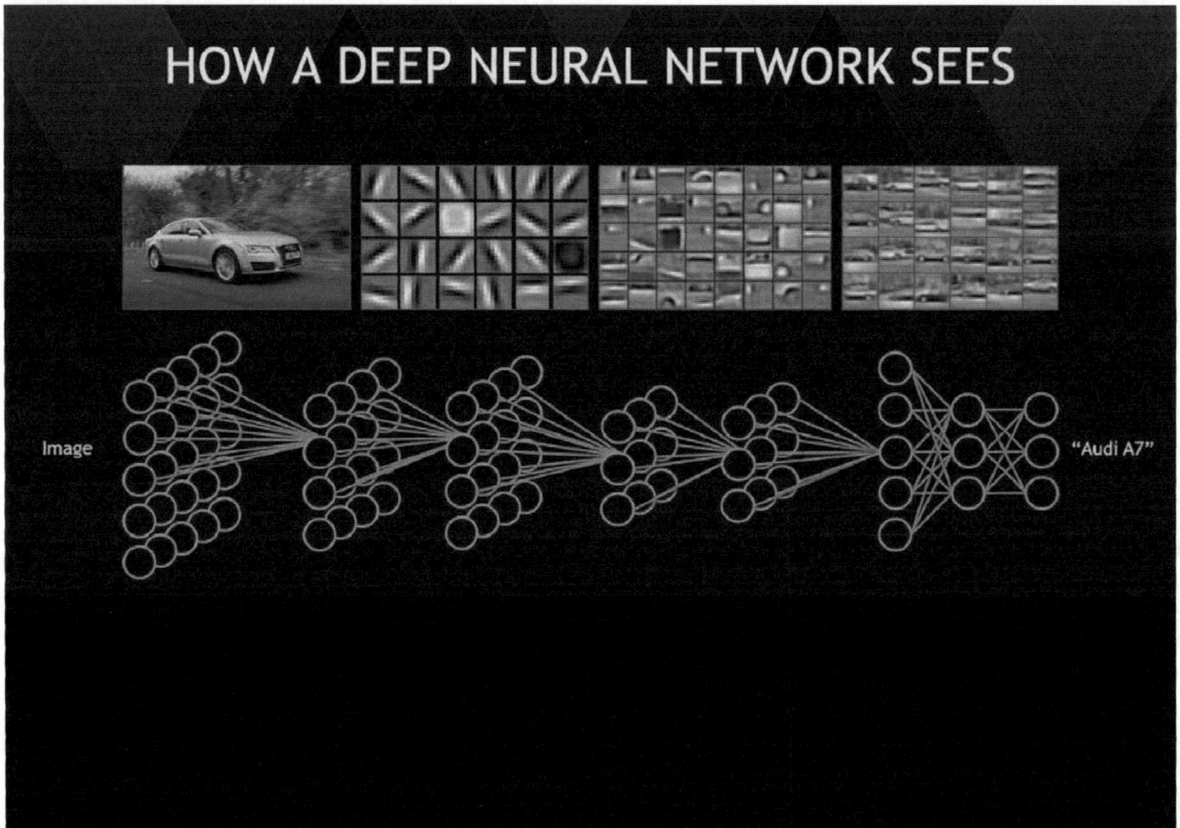
HOW A DEEP NEURAL NETWORK SEES
Image
"Audi A7"
Image source: "Unsupervised Learning of Hierarchical Representations with Convolutional Deep Belief Networks" ICML 2009 & Comm. ACM 2011.
Honglak Lee, Roger Grosse, Rajesh Ranganath, and Andrew Ng.

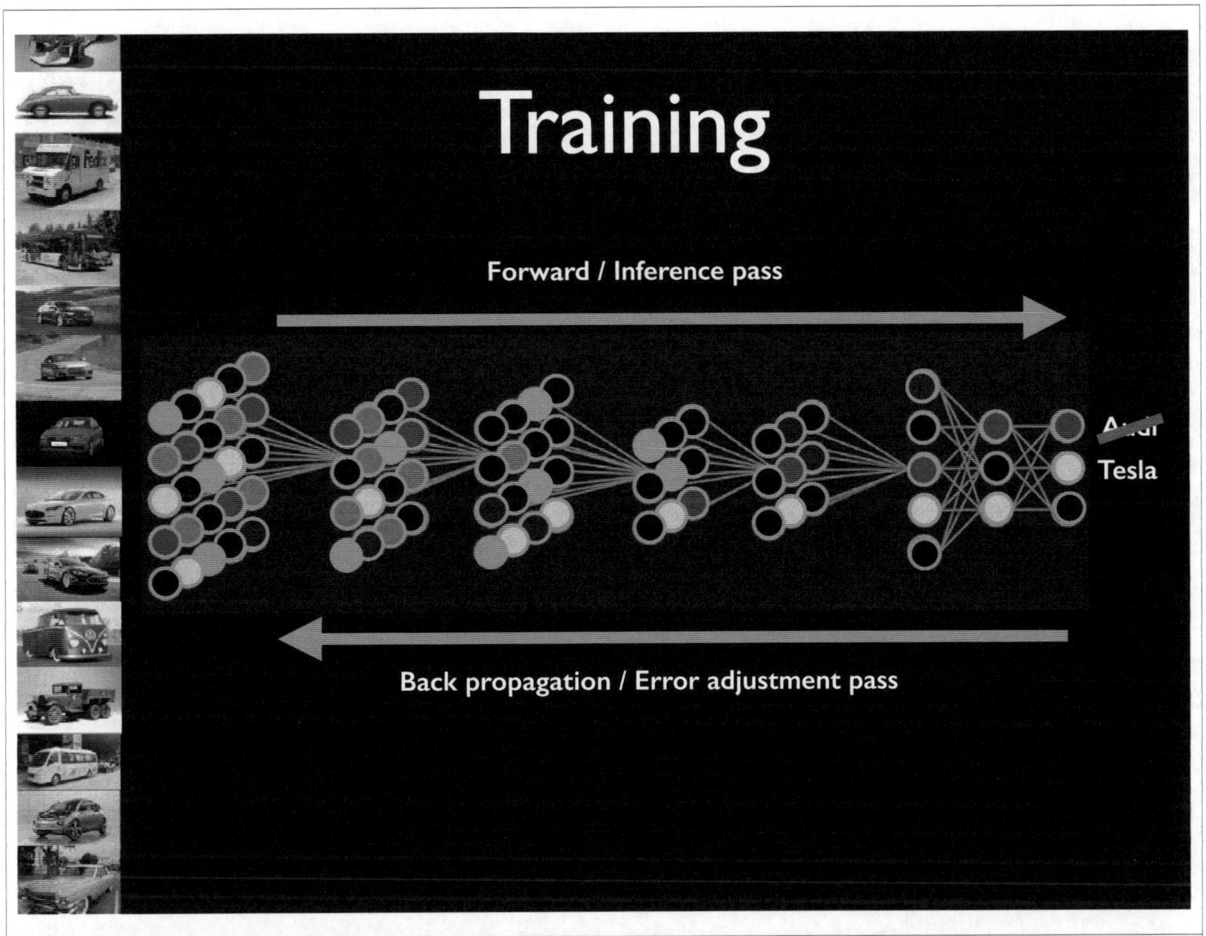
Training
Forward / Inference pass
Tesla
Back propagation / Error adjustment pass

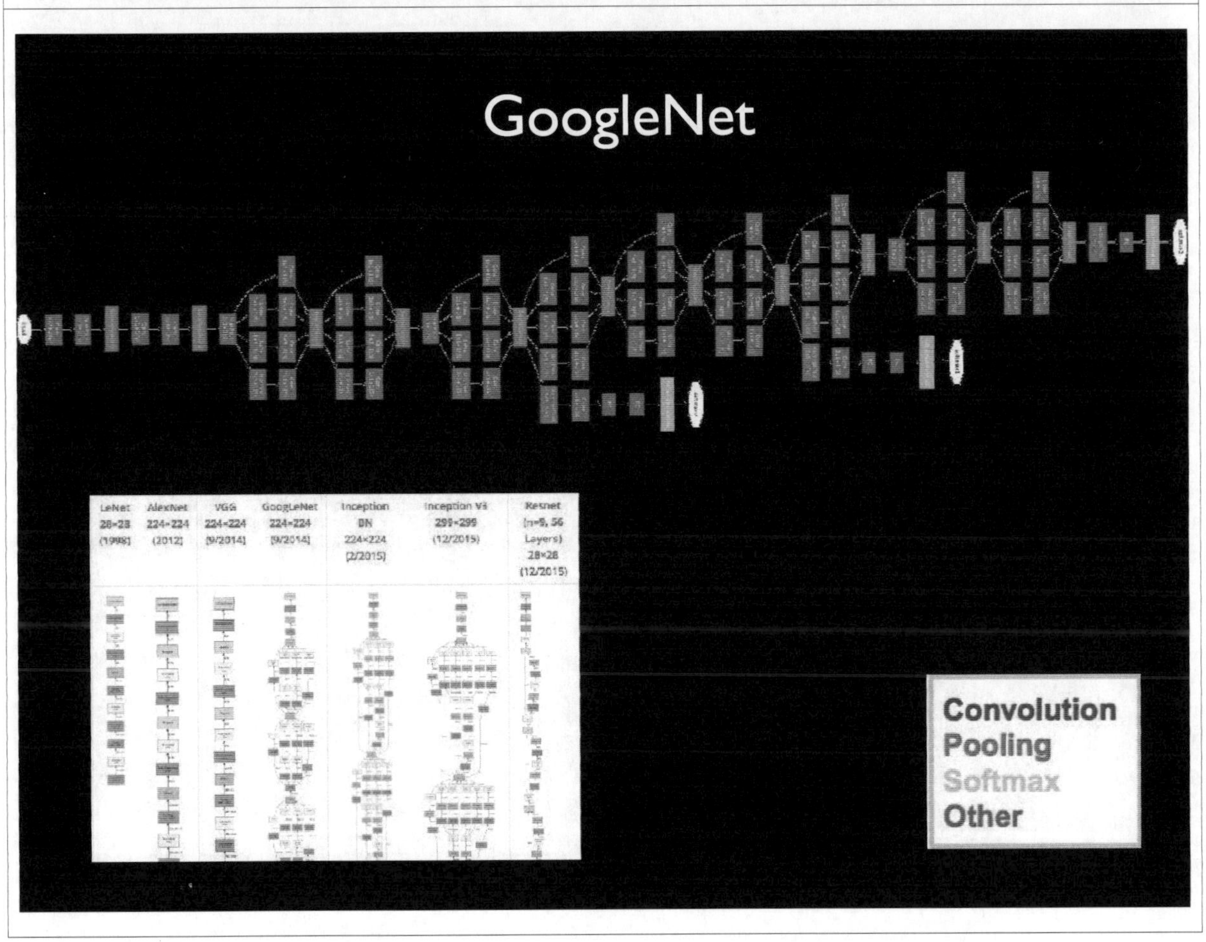
GoogleNet
LeNet 28×28 (1998)
AlexNet 224×224 (2012)
VGG 224×224 (9/2014)
GoogLeNet 224×224 (9/2014)
Inception BN 224×224 (2/2015)
Inception V3 299×299 (12/2015)
Resnet (n=5, 56 Layers) 28×28 (12/2015)
Convolution
Pooling
Softmax
Other

AV Architecture

Sense → Plan → Actuate

Sense
- Sensors & fusion
- Object Detection & tracking
- Behavior Detection
- Failure detection

- camera, lidar, radar, ultrasound
- V2x, 5G
- GPS, IMU, map/localization
- lead vehicle

Plan
- Trajectory control
- Maneuver planning
- e-horizon, maps
- Traffic signs, rules, lights
- Safe distances
- Comfortable dynamics

Actuate
- steering
- brakes, accelerator
- belt tighteners
- ECUs

End-to-End learning

Sense → Actuate

Raw sensor inputs

steering, brake outputs

http://www.cs.nyu.edu/~yann/research/dave/

14 years ago
Urs Muller, Yann LeCun
Training data: human driver,
100K images + stick input
72K nodes, 3M connections
no hand-coded algorithms

End-to-End learning

NVidia BB8 car running PilotNet

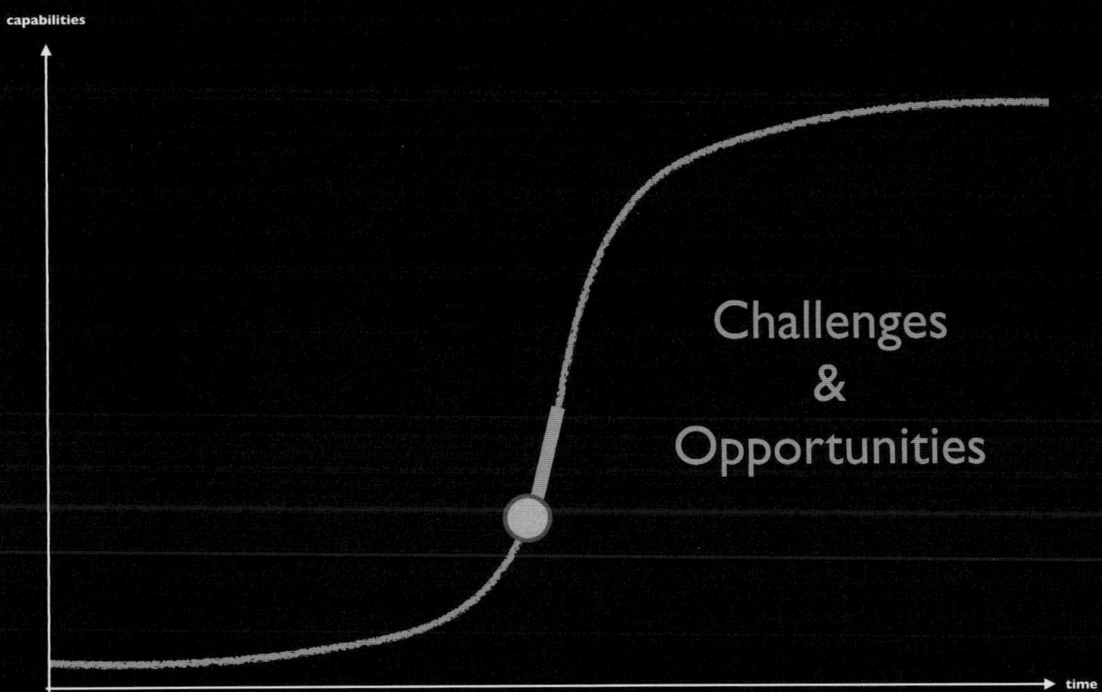

Challenges: Introspection & Development Tools

Engineering "V"

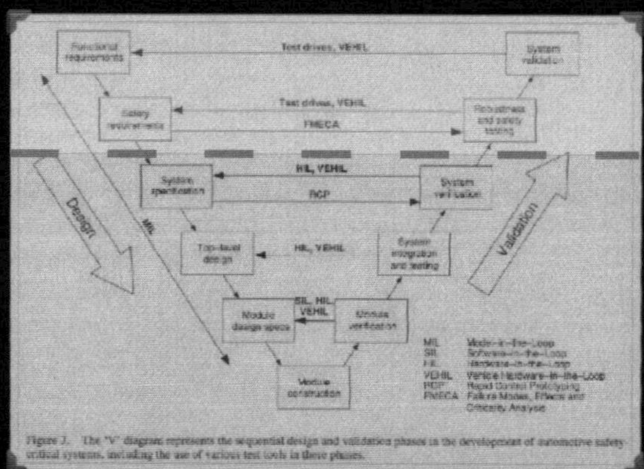

Figure 3. The "V" diagram represents the sequential design and validation phases in the development of automotive safety critical systems, including the use of various test tools in these phases.

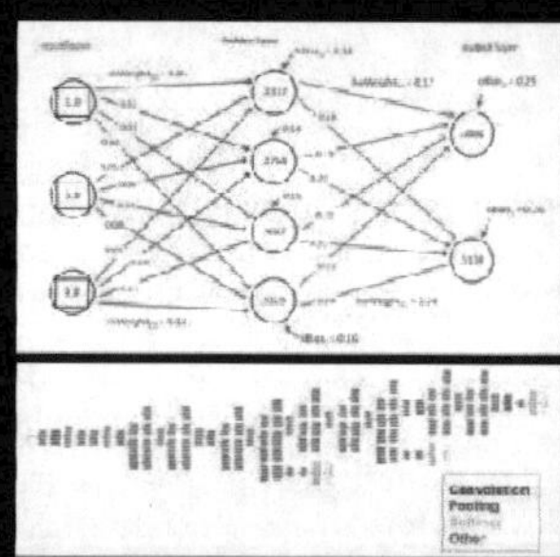

How does Deep Learning map to industry standard processes, ISO 26262 etc. ?

Modern Coding Tools+Methods

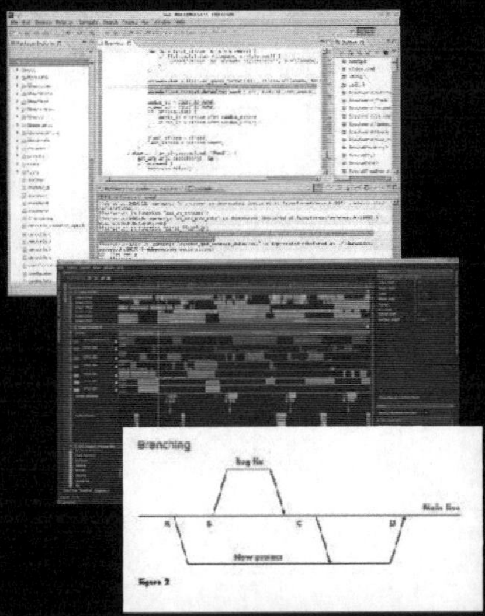

- Editor
- Debugger
- Profiler
- Static code analysis
- Versioning
- Testing
- Code coverage
- Requirements tracking

- Agile, SCRUM, CI, TDD

NN: still more art than science

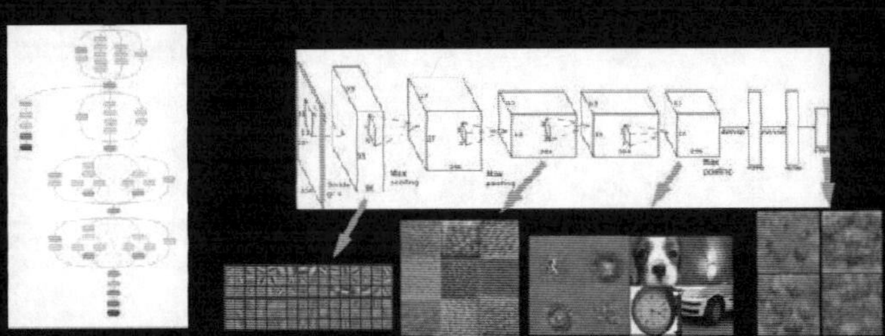

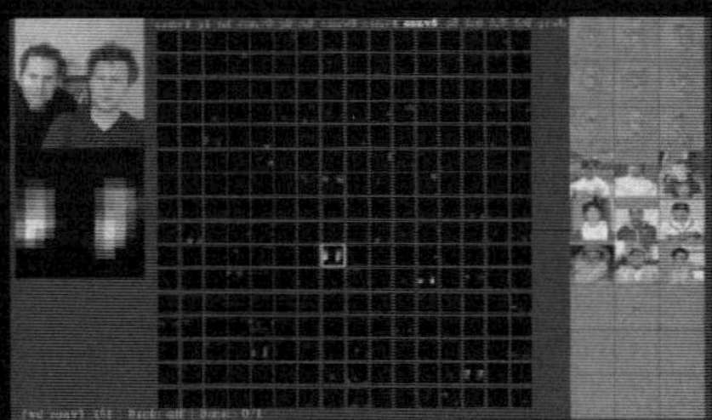

Deep Visualization Toolbox

- Network architecture
  - Height / Width / Reusable modules
- Training set size, backprop f's
- Don't run over grandma
- Adding outputs w/o retraining
- Activation semantics
- Undertrained parts

Spatial Saliency visualization (NVidia)

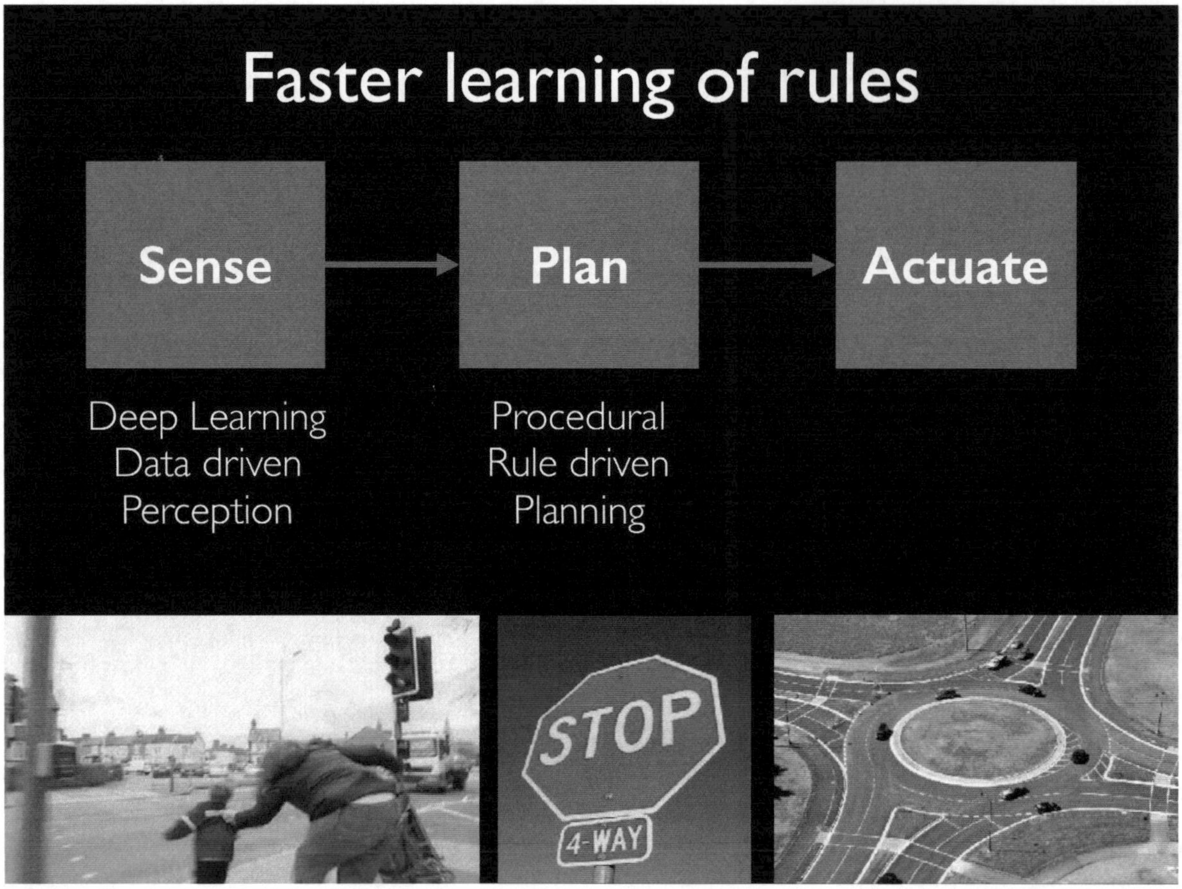
Faster learning of rules
Sense
Plan
Actuate
Deep Learning
Data driven
Perception
Procedural
Rule driven
Planning
STOP
4-WAY

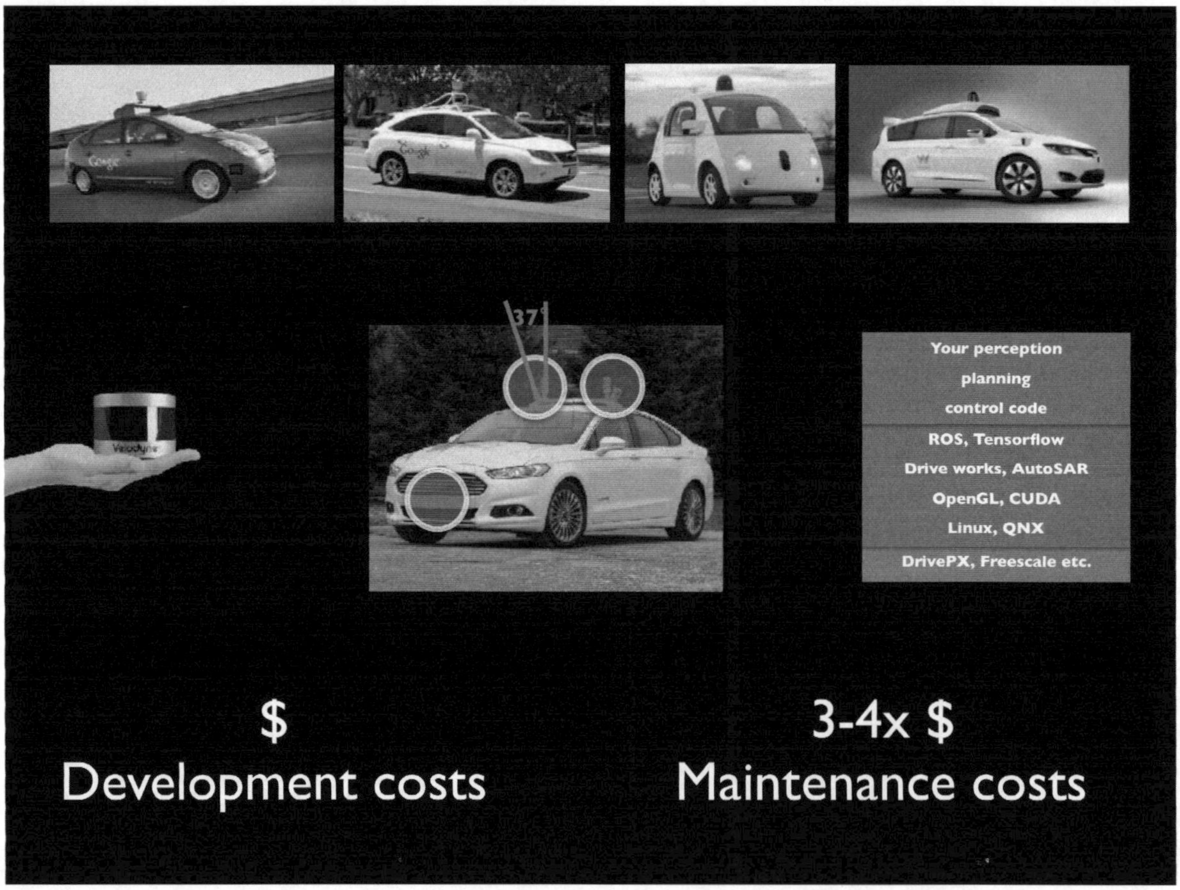
Your perception
planning
control code
ROS, Tensorflow
Drive works, AutoSAR
OpenGL, CUDA
Linux, QNX
DrivePX, Freescale etc.
$
Development costs
3-4x $
Maintenance costs

Opportunities

Functional & Safety validation

using

Sensor Simulation

Autoliv
Autoliv
30 TESTS / DAY

Cityscapes dataset
20K images hand labeled data, 30 classes, 50 cities, Daimler/Max Planck/TU Darmstadt
https://www.cityscapes-dataset.com/

Tesla collecting driver behavior

>100K vehicles, >3M miles per day

The world is way too big to just *randomly* sample it.

Phillip Isola (MIT/UC Berkeley)

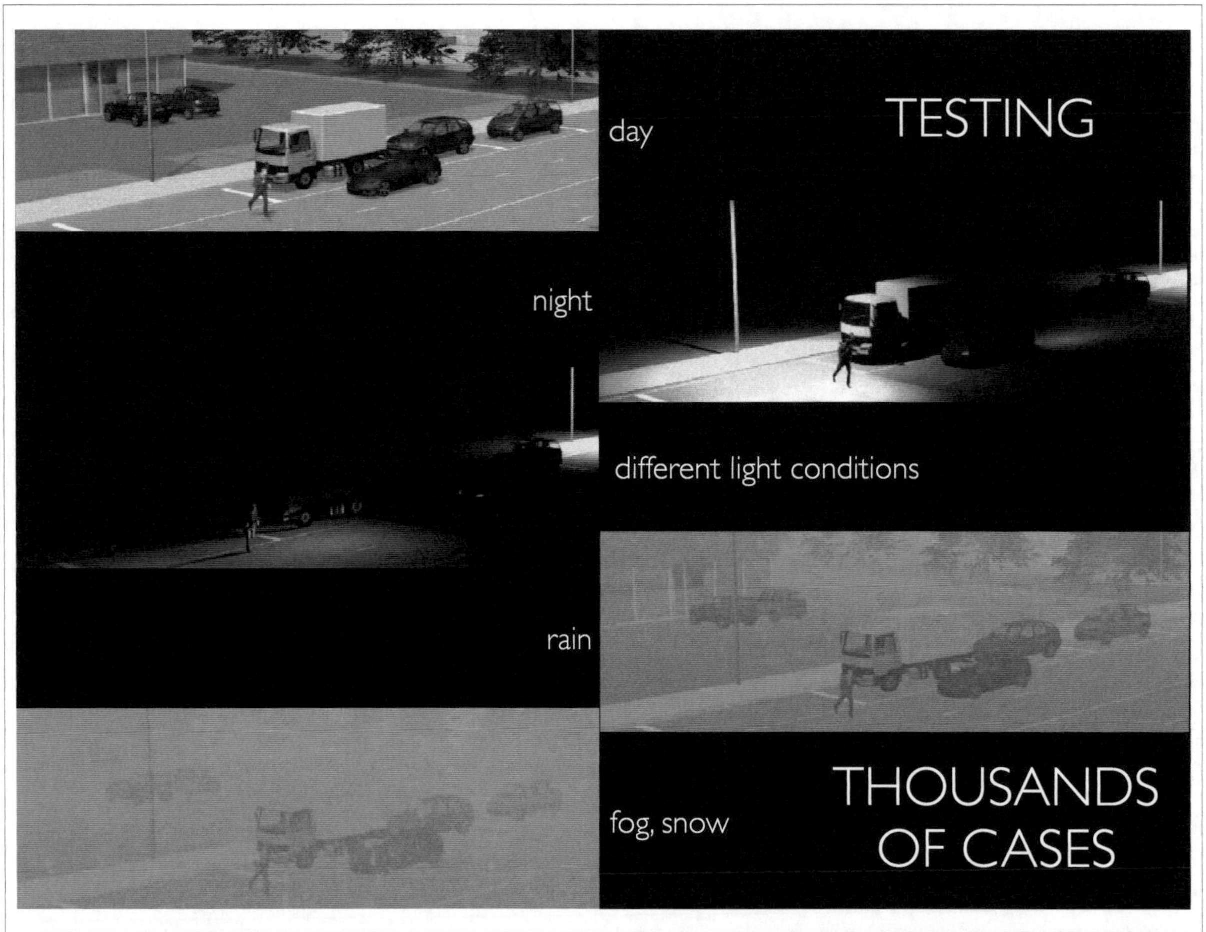
TESTING
day
night
different light conditions
rain
fog, snow
THOUSANDS
OF CASES

3 MILLION
TESTS / DAY

Simulation: a key driver for AV success

- Controllable (weather, lighting, sensor configs etc.)
- Repeatable, fuzz-able
- Scalable
- Dangerous = ok
- Lots of training data, fully labeled, with ground truth
- **Sim Training is transferable**

Sensor Fusion

038402
698 TSW
Scirocco.
V131 K

RELEVANT OBJECTS
HIDDEN BY TREES OR OTHER CARS
CAN STILL BE SEEN AND ANTICIPATED
Civil Maps

Sensor Steering

foveation requires loops rather than just feedforward

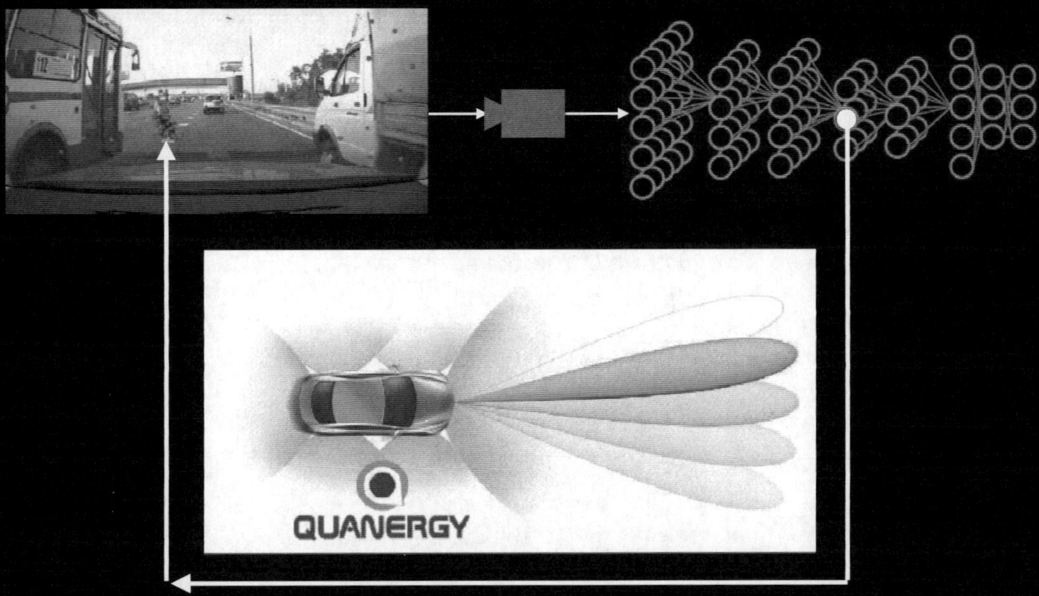

Behavior
not just
Object recognition

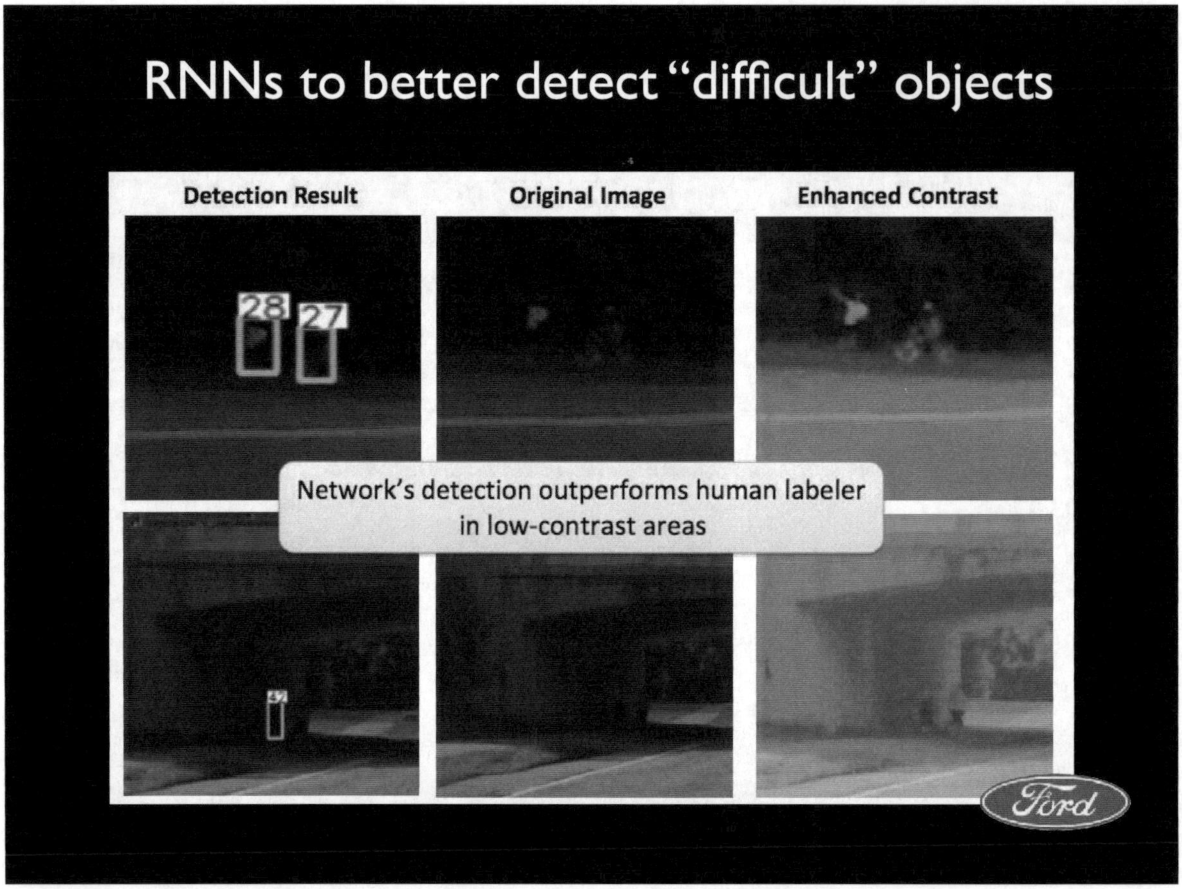
RNNs to better detect "difficult" objects
Detection Result
Original Image
Enhanced Contrast
28
27
Network's detection outperforms human labeler in low-contrast areas
Ford

Behavior classification

Eye contact
Head pose, gaze estimation
THE EASIEST WAY TO AVOID A COLLISION IS TO ALWAYS MAKE EYE CONTACT WITH DRIVERS.
PEDESTRIAN SAFETY

Intent-Aware Long-Term Prediction of Pedestrian Motion
http://vision.ucla.edu/papers/karasevAHS16.pdf

Sharing Data & Best Practices

- Scenario catalog
  - scenarios (geometry, traffic, environment etc.)
  - eg freeway maneuvers
  - German project:
- Accident Data
- Realistic traffic & pedestrian data

PEGASUS

http://www.pegasus-projekt.info/

This has only just begun

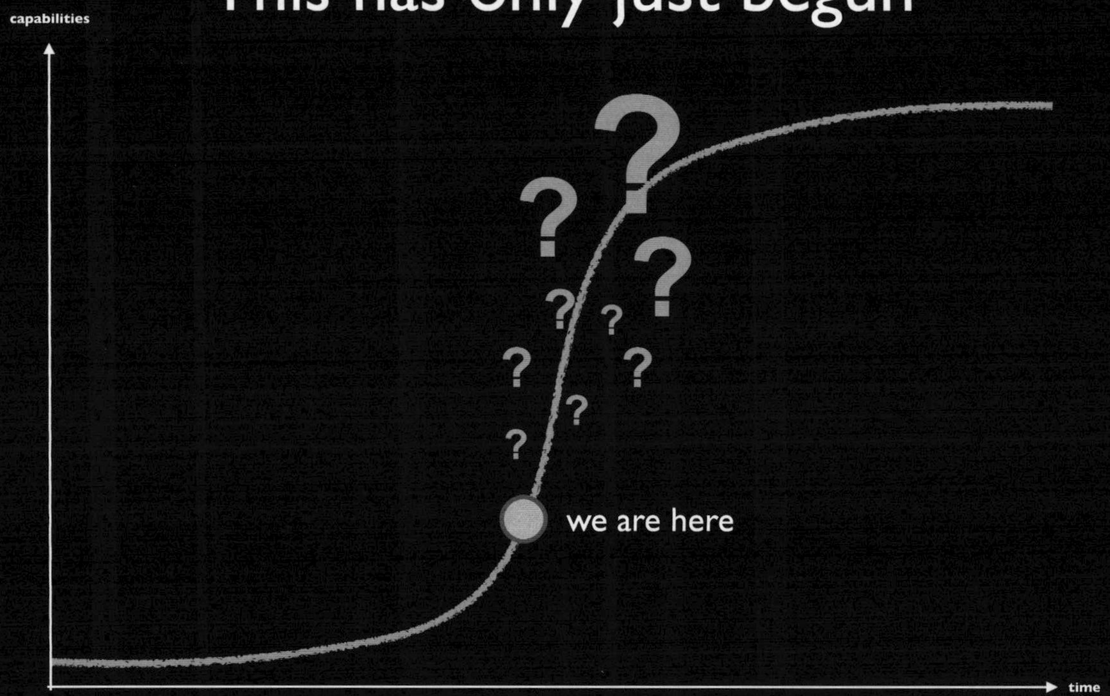

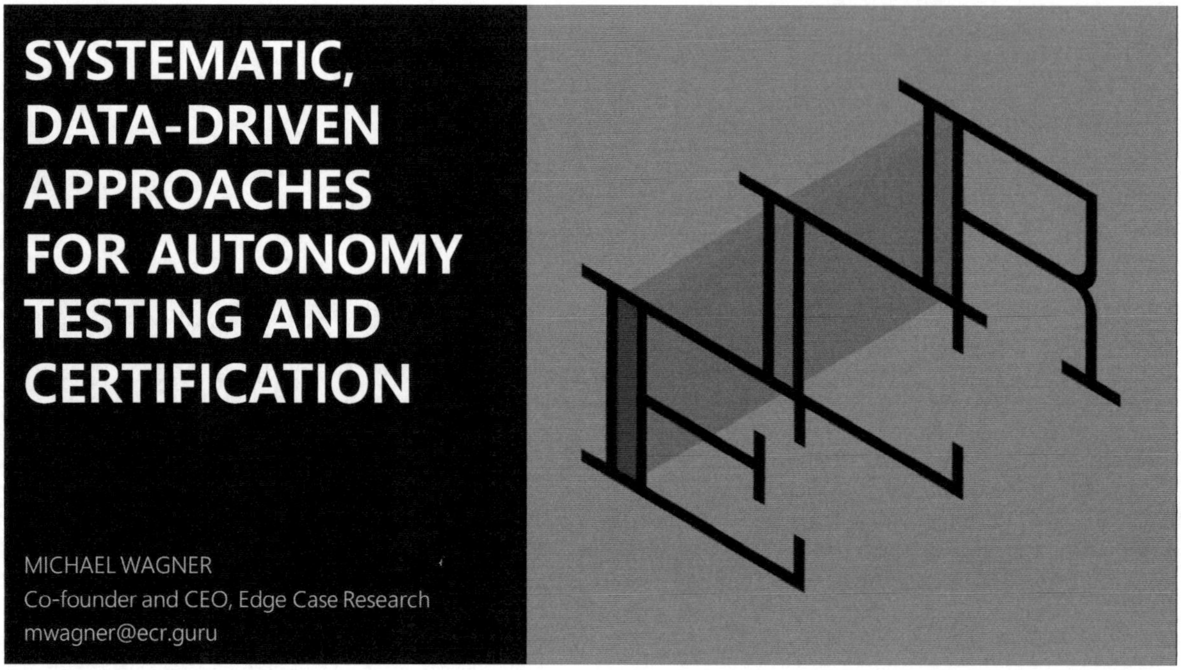
SYSTEMATIC, DATA-DRIVEN APPROACHES FOR AUTONOMY TESTING AND CERTIFICATION
MICHAEL WAGNER
Co-founder and CEO, Edge Case Research
mwagner@ecr.guru

IT'S TIME TO DRAFT SOME GUIDELINES FOR BUILDING SAFER AUTONOMY
MICHAEL WAGNER
Co-founder and CEO, Edge Case Research
mwagner@ecr.guru

WHAT "SAFE SOFTWARE" MEANS

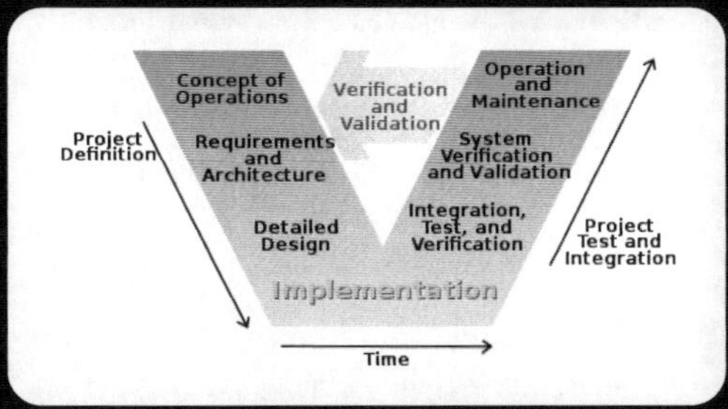

+

JOINT SERVICES – SOFTWARE SAFETY AUTHORITIES (JS-SSA)

Software System Safety

Implementation Process and Tasks Supporting MIL-STD-882E

Developed by the JS-SSA Sub-Working Group in Support of the Joint Services Weapon Safety Review (JSWSR) Boards

April 1, 2016

We claim that we've identified all the risks,
Our safety specification mitigates the risks we've identified,
The system satisfies our safety specification, and ...

... that we built the system according to the best-practices of our industry.

www.ecr.guru

SO WHAT'S DIFFICULT ABOUT AUTONOMY?

Arguing that autonomy will be safe in all unspecified environments demands unbounded validation!

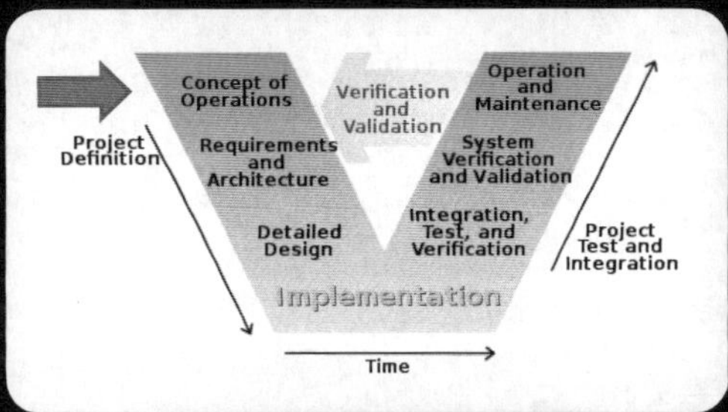

www.ecr.guru

SO WHAT'S DIFFICULT ABOUT AUTONOMY?

Arguing that autonomy will be safe in all unspecified environments demands unbounded validation!

So we define operational design domains that constrain the situations in which autonomy must behave safely.

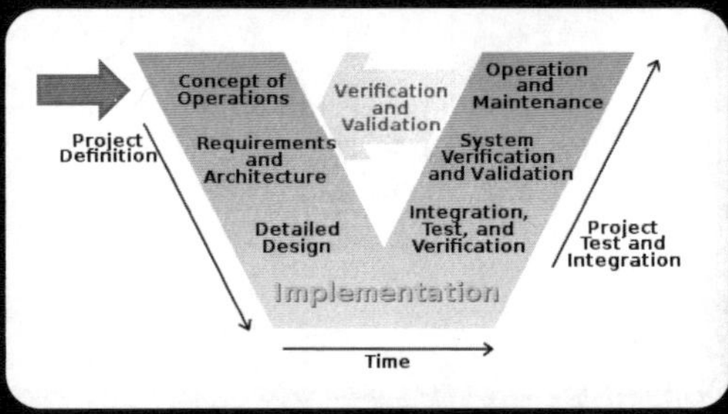

www.ecr.guru

SO WHAT'S DIFFICULT ABOUT AUTONOMY?

Arguing that autonomy will be safe in all unspecified environments demands unbounded validation!

So we define operational design domains that constrain the situations in which autonomy must behave safely.

Unfortunately, domains that seem equivalent to humans may not be equivalent to autonomy.

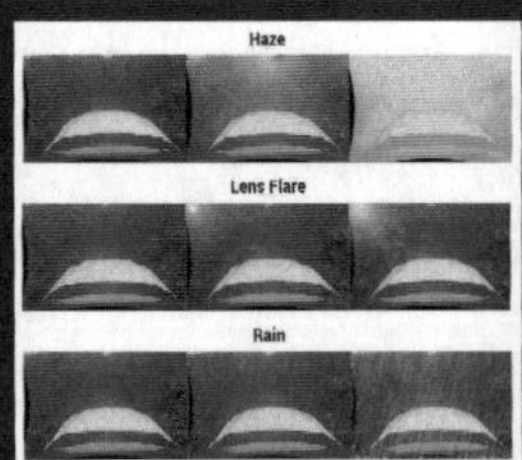

Exploring the response of a DNN to environmental perturbations from "Robustness Testing for Perception Systems", RIOT Project, Carnegie Mellon University. Figures are DISTRIBUTION A.

www.ecr.guru

THE IMPACT OF SMALL CHANGES CAN BE SURPRISING

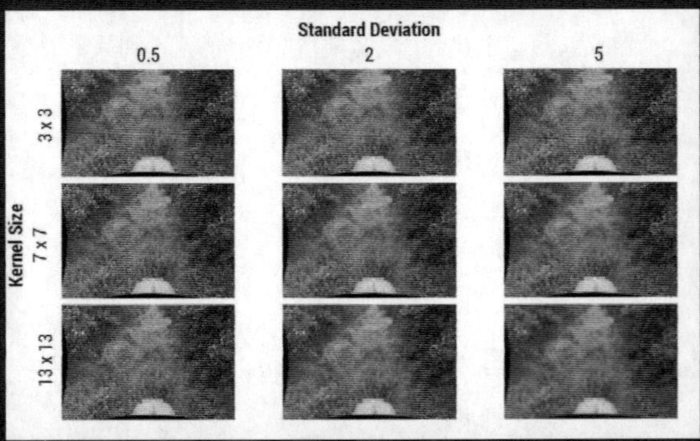

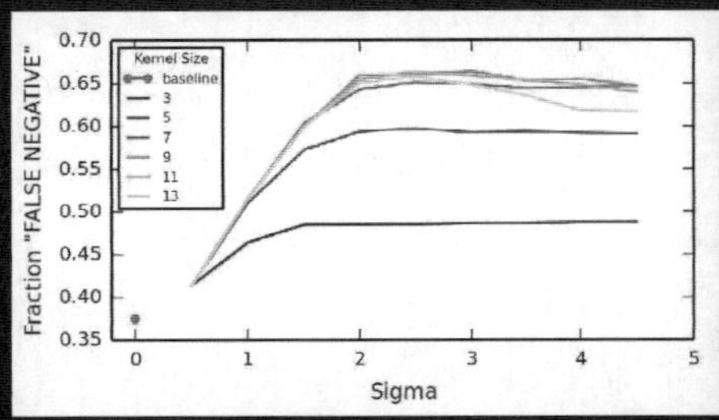

Correct detection

False negative

www.ecr.guru

THE IMPACT OF SMALL CHANGES CAN BE SURPRISING

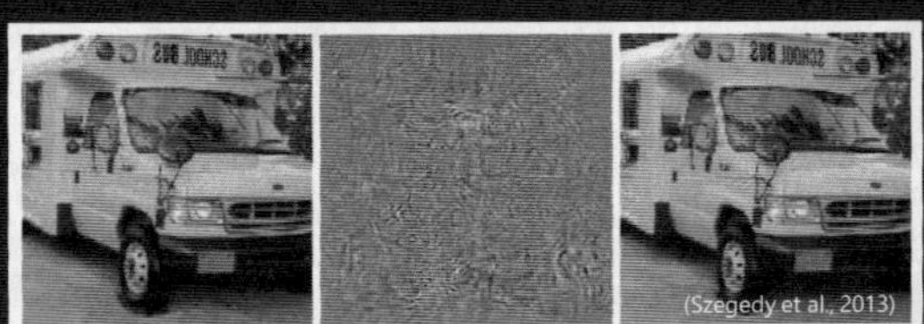

Bus Magnified Difference Not a Bus

Car Not a Car Magnified Difference

www.ecr.guru

AND WHAT ABOUT REQUIREMENTS ON DETECTING OBJECTS?

"Stop when there is a pedestrian in a crosswalk."

www.ecr.guru

MACHINE LEARNING DOESN'T HAVE REQUIREMENTS

We rarely have explicit definitions of the objects that the AV needs to respond to. We turn to machine learning, which uses no requirements. But we no longer have traceability.

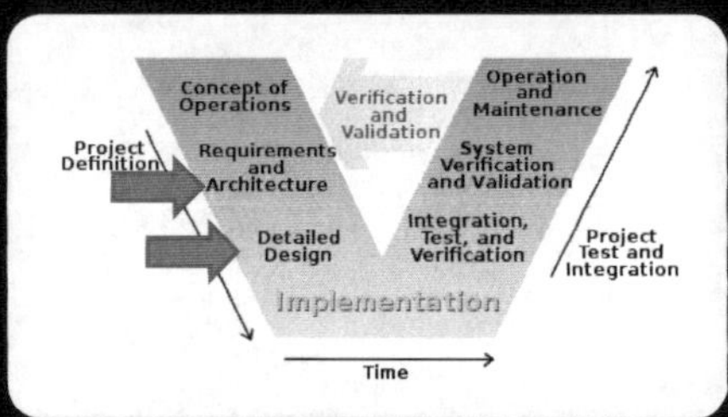

www.ecr.guru

SOME USEFUL IDEAS: SIMPLEX ARCHITECTURE PATTERN

A simple, verified Safety Controller can help mitigate risks in an unverified Complex Controller. DARPA HACMS program used it to address security risks.

The Decision Logic module is responsible for swapping in the Safety Controller **before** an accident is unavoidable.

Simplex approach is useful for supporting phased-deployment safety concepts.

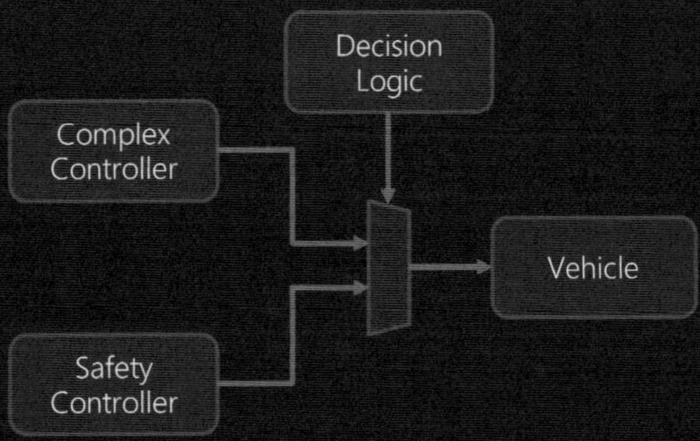

www.ecr.guru

MORE ADVANCED ARCHITECTURAL PATTERNS

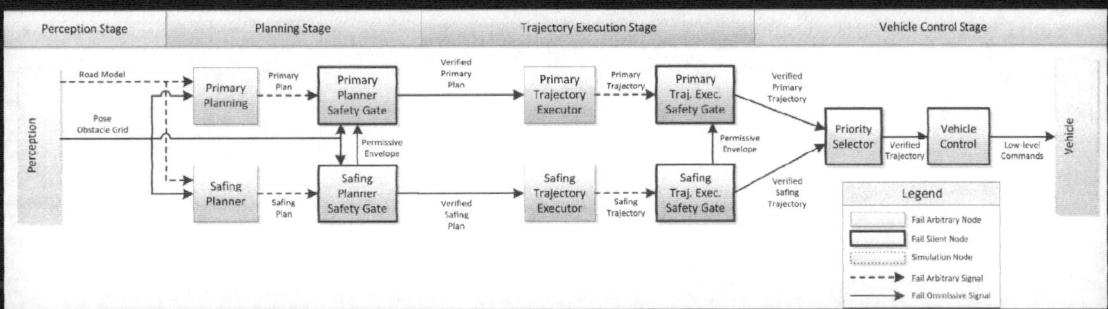

Our team at Carnegie Mellon designed and submitted a patent for a two-channel architecture shown here, that segments autonomy architectures into checkable pipeline stages, and defines safing behaviors that are triggered when a primary-channel safety gate fails silently. (Figure is Distribution A, NREC Case #: 2016-01-19.)

www.ecr.guru

SOME USEFUL IDEAS: EVOLVING SAFETY CONCEPTS

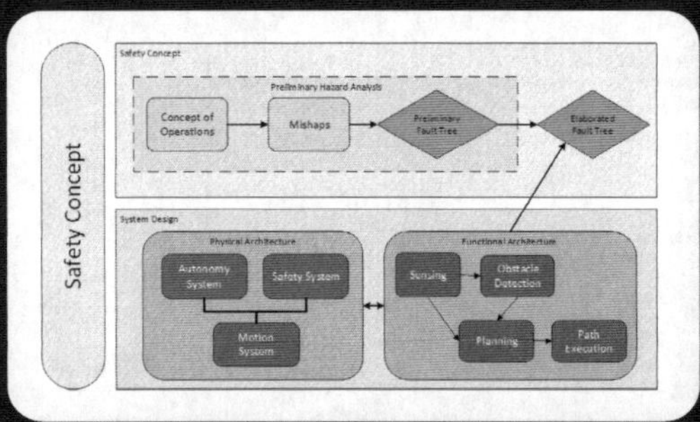

ECR's model-based iterative safety analysis (MISSA)

Model the operational domains and the function architecture of the AV to generate a fault tree.

Use the fault tree to identify risks and opportunities for mitigations.

Iterate safety concept to deploy more sophisticated technology in ever more complex operational domains (Schladover et al., 2001), (Bayouth and Koopman, 1998).

www.ecr.guru

BUT WHAT ABOUT PERCEPTION?

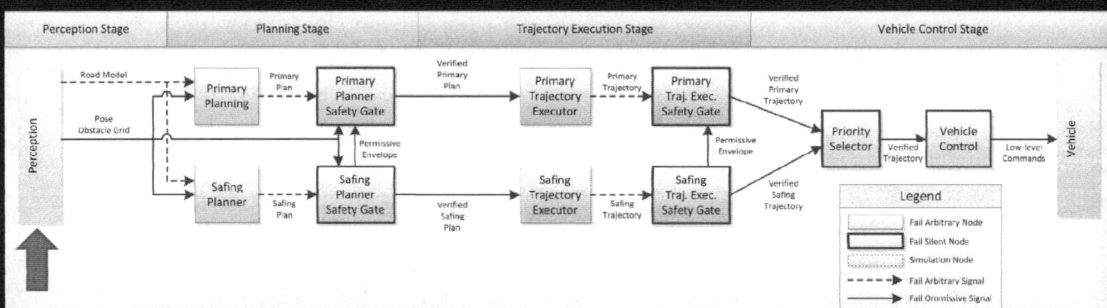

www.ecr.guru

DEEP LEARNING: EXPLAINABILITY AND VERIFICATION

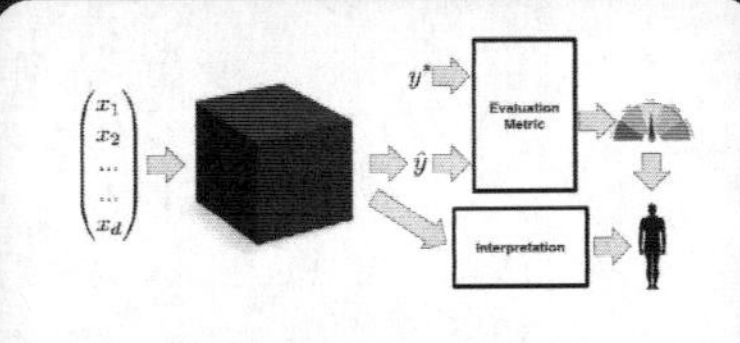

Figure 1. Typically, evaluation metrics require only predictions and *ground truth* labels. When stakeholders additionally demand *interpretability*, we might infer the existence of desiderata that cannot be captured in this fashion.

(Lipton, 2016)

"Baseball"
(Nguyen et al., 2015)

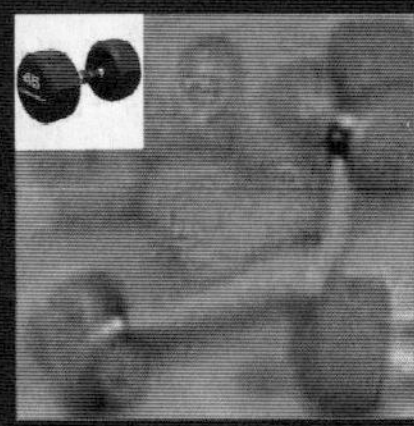

"Dumbbell"
(Mordvintsev et al., 2015)

"The associations learned by supervised learning algorithms are not guaranteed to reflect causal relationships."

Verification is different than explainability. It is the **independent determination** as to whether the necessary casual relationships are in place.

www.ecr.guru

SOME USEFUL IDEAS: EXPERIMENTAL EVALUATION

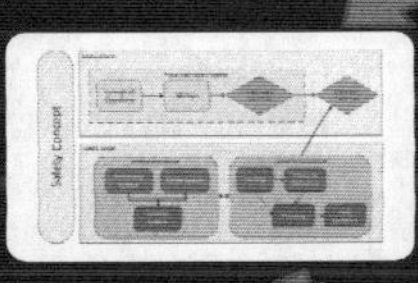

An evolving safety concept

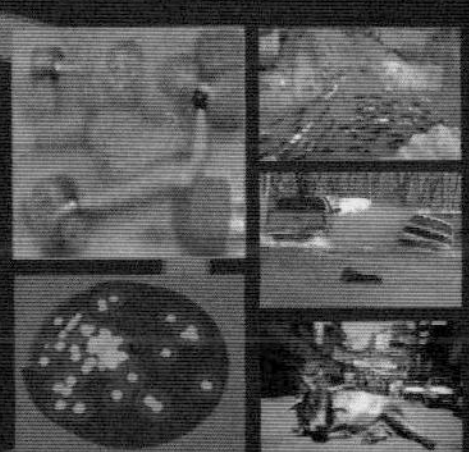

Rigorous experiments to evaluate causality

"Shift from the simple, binary view of standards to a staged, dynamic evaluation resembling the approval process for drugs" **(which also happen to be inscrutable black boxes)**. Design **controlled experiments** on the deep learner to:

- Generate data about the actual safe operational domain, and
- Find suitable spaces in which we can compare the similarity between given operating environments.

Successful "early trials" permit ever more expanded operational domains.

This isn't just testing to rack up miles, these would be scientific and reviewable experimental programs.

www.ecr.guru

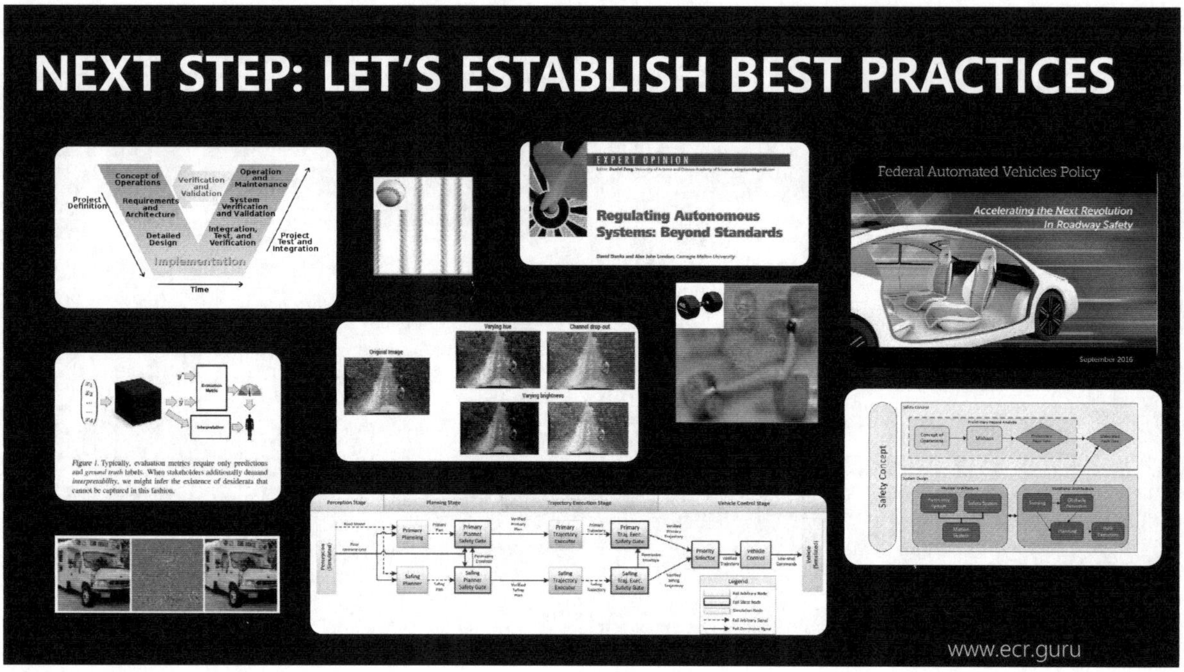
NEXT STEP: LET'S ESTABLISH BEST PRACTICES
Concept of Operations
Verification and Validation
Operation and Maintenance
Project Definition
Requirements and Architecture
System Verification and Validation
Detailed Design
Integration, Test, and Verification
Project Test and Integration
Implementation
Time
EXPERT OPINION
Regulating Autonomous Systems: Beyond Standards
Federal Automated Vehicles Policy
Accelerating the Next Revolution In Roadway Safety
September 2016
Original image
Varying hue
Channel drop-out
Varying brightness
Figure 1. Typically, evaluation metrics require only predictions and ground truth labels. When stakeholders additionally demand interpretability, we might infer the existence of desiderata that cannot be captured in this fashion.
Safety Concept
Perception Stage
Planning Stage
Trajectory Execution Stage
Vehicle Control Stage
Primary Planning
Primary Planner Safety Gate
Primary Trajectory Executor
Primary Traj. Exec. Safety Gate
Safing Planner
Safing Planner Safety Gate
Safing Trajectory Executor
Safing Traj. Exec. Safety Gate
Priority Selector
Vehicle Control
Legend
www.ecr.guru

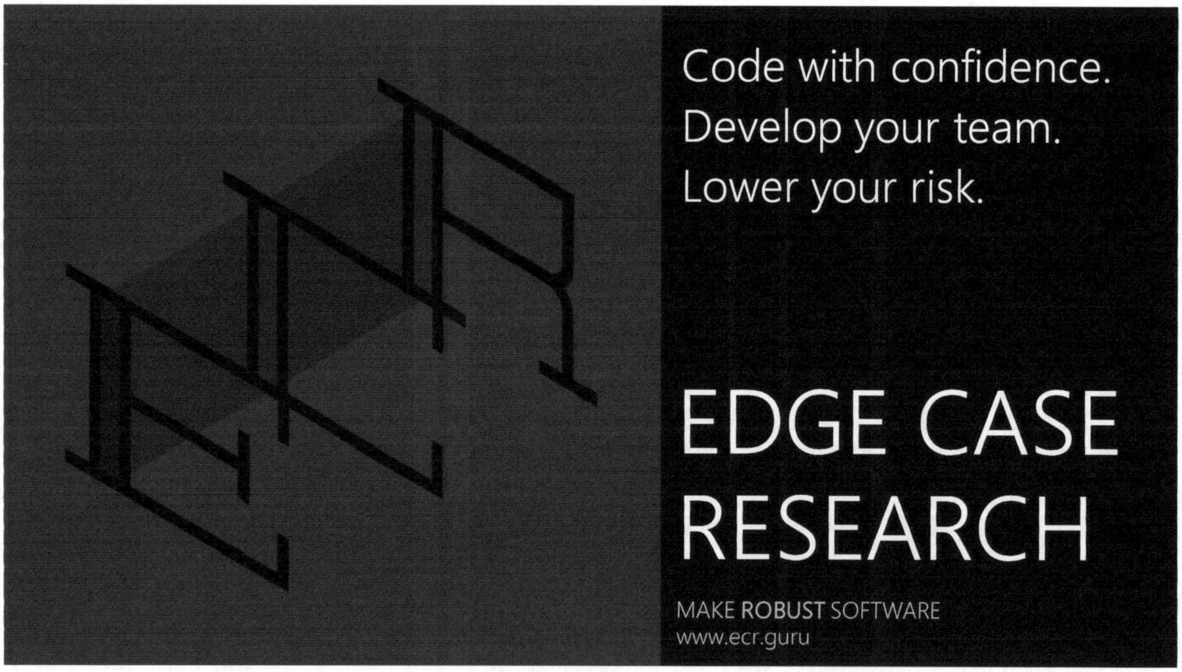
Code with confidence.
Develop your team.
Lower your risk.
EDGE CASE RESEARCH
MAKE ROBUST SOFTWARE
www.ecr.guru

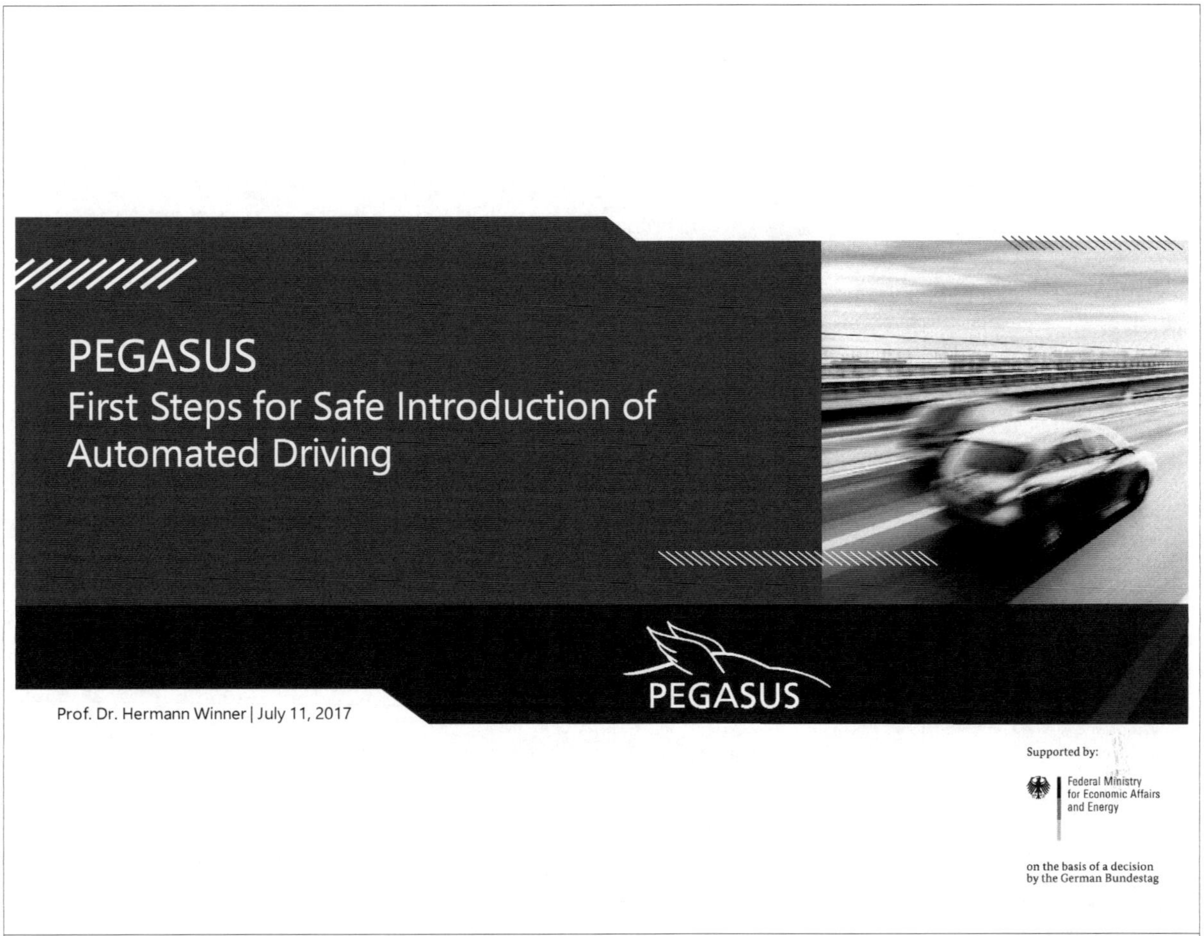
PEGASUS
First Steps for Safe Introduction of Automated Driving
PEGASUS
Prof. Dr. Hermann Winner | July 11, 2017
Supported by:
Federal Ministry for Economic Affairs and Energy
on the basis of a decision by the German Bundestag

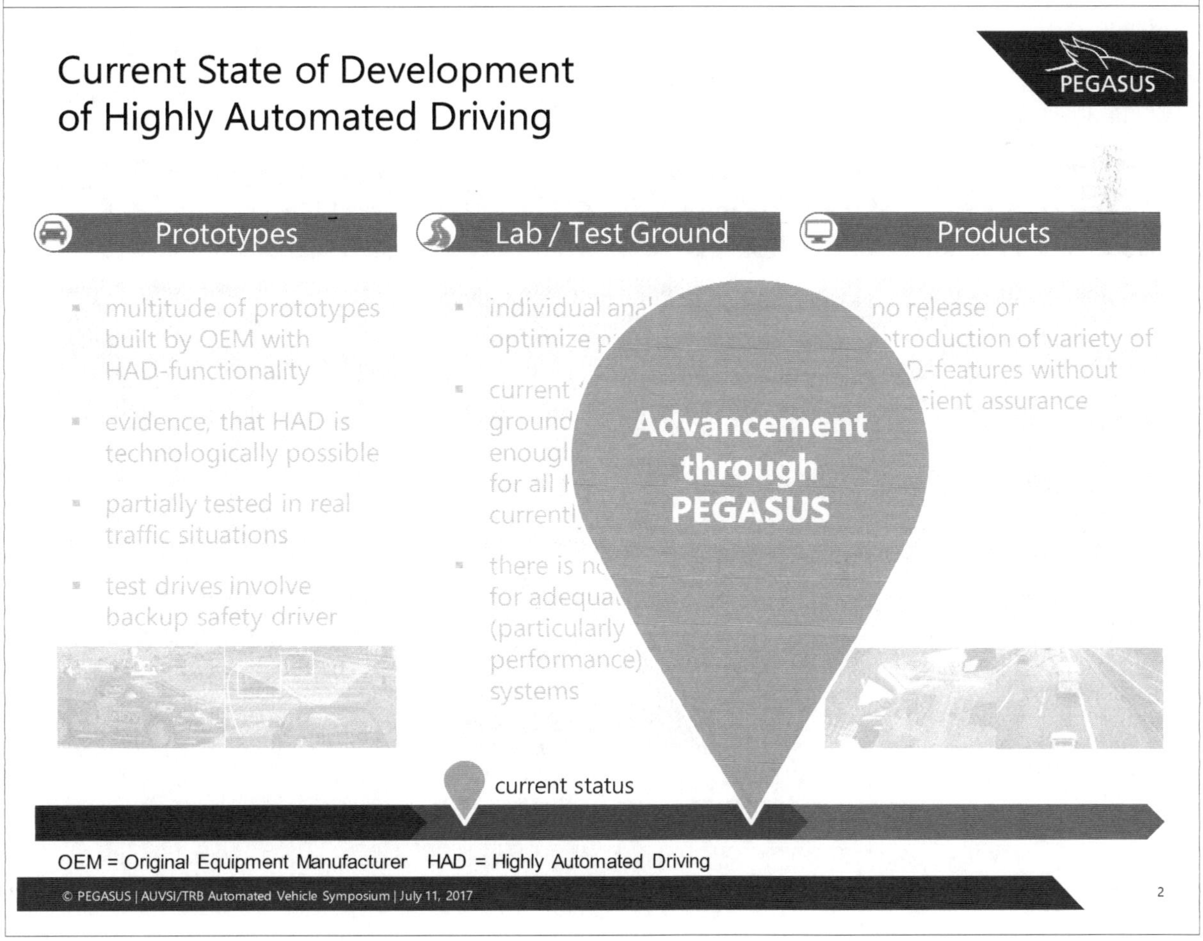
Current State of Development of Highly Automated Driving
PEGASUS
Prototypes
Lab / Test Ground
Products
multitude of prototypes built by OEM with HAD-functionality
evidence, that HAD is technologically possible
partially tested in real traffic situations
test drives involve backup safety driver
Advancement through PEGASUS
current status
OEM = Original Equipment Manufacturer HAD = Highly Automated Driving
© PEGASUS | AUVSI/TRB Automated Vehicle Symposium | July 11, 2017
2

PEGASUS Key Figures

PEGASUS

project for the **e**stablishment of **g**enerally **a**ccepted quality criteria, tools and methods as well as **s**cenarios and **s**ituations for the release of highly-automated driving functions

42 months term	January 2016 – June 2019
17 partners	▪ OEM: Audi, BMW, Daimler, Opel, Volkswagen ▪ Tier 1: ADC Automotive Distance Control, Bosch, Continental Teves ▪ Test Lab: TÜV SÜD ▪ SMB: fka, iMAR, IPG, QTronic, TraceTronic, VIRES ▪ scientific institutes: DLR, TU Darmstadt
Affiliated partners & Subcontracts	▪ i.a. BASt, IFR, ika, OFFIS
Project volume	▪ approx. 34,5 Mio. EUR ▪ subsidies: 16,3 Mio. EUR
Personnel deployment	▪ approx. 1.791 man-month or 149 man-years

Project coordination, Project office

Central Issues of the PEGASUS Project

PEGASUS

What level of performance is expected of an automated vehicle?
How can we verify that it achieves the desired performance consistently?

Scenario Analysis & Quality Measures	Implementation Process	Testing	Reflection of Results & Embedding
What human capacity does the application require? What about technical capacity? Is it sufficiently accepted? Which criteria and measures can be deducted from it?	Which tools, methods and processes are necessary?	How can completeness of relevant test runs be ensured? What do the criteria and measures for these test runs look like? What can be tested in labs or in simulation? What must be tested on test grounds, what must be tested on the road?	Is the concept sustainable? How does the process of embedding work?

Central Issues of the PEGASUS Project

PEGASUS

What level of performance is expected of an automated vehicle?
How can we verify that it achieves the desired performance consistently?

Scenario Analysis & Quality Measures	Implementation Process	Testing	Reflection of Results & Embedding

What human capacity does the application require?

What about technical capacity?

Is it sufficiently accepted?

Which criteria and measures can be deducted from it?

Scenario Analysis and Quality Measures

PEGASUS

Deduction of requirements based on the accepted measure of quality

Deduction of an **accepted quality measure** for automated driving features

Determination of safety level through assessment of probability of occurrence and mechanical manageability in critical situations

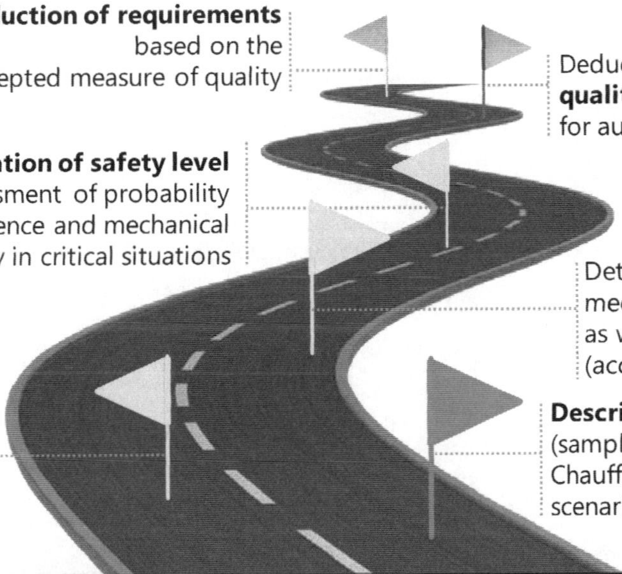

Determination of human and mechanical **performance** as well as **effectiveness** (accident avoidance potential)

Determination of critical **traffic situations**

Description of application scenario (sample application: Highway Chauffeur + enhanced application scenario)

PEGASUS Metrics Perspective

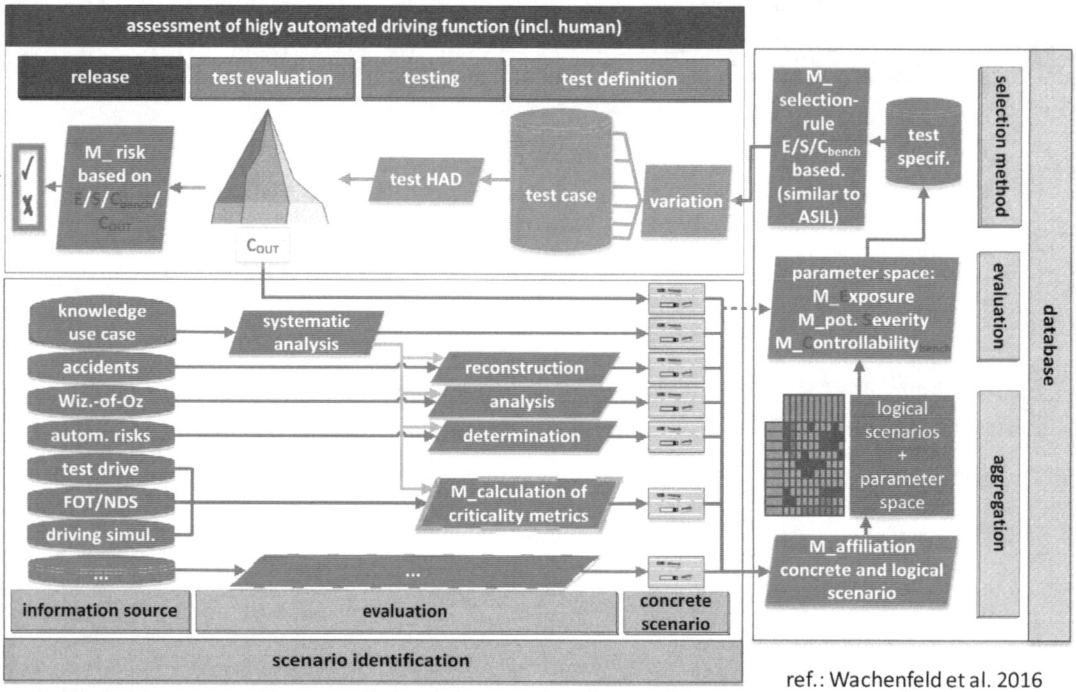

ref.: Wachenfeld et al. 2016

Central Issues of the PEGASUS Project

What level of performance is expected of an automated vehicle?
How can we verify that it achieves the desired performance consistently?

Scenario Analysis & Quality Measures

Implementation Process

Which tools, methods and processes are necessary?

Testing

Reflection of Results & Embedding

Implementation Process

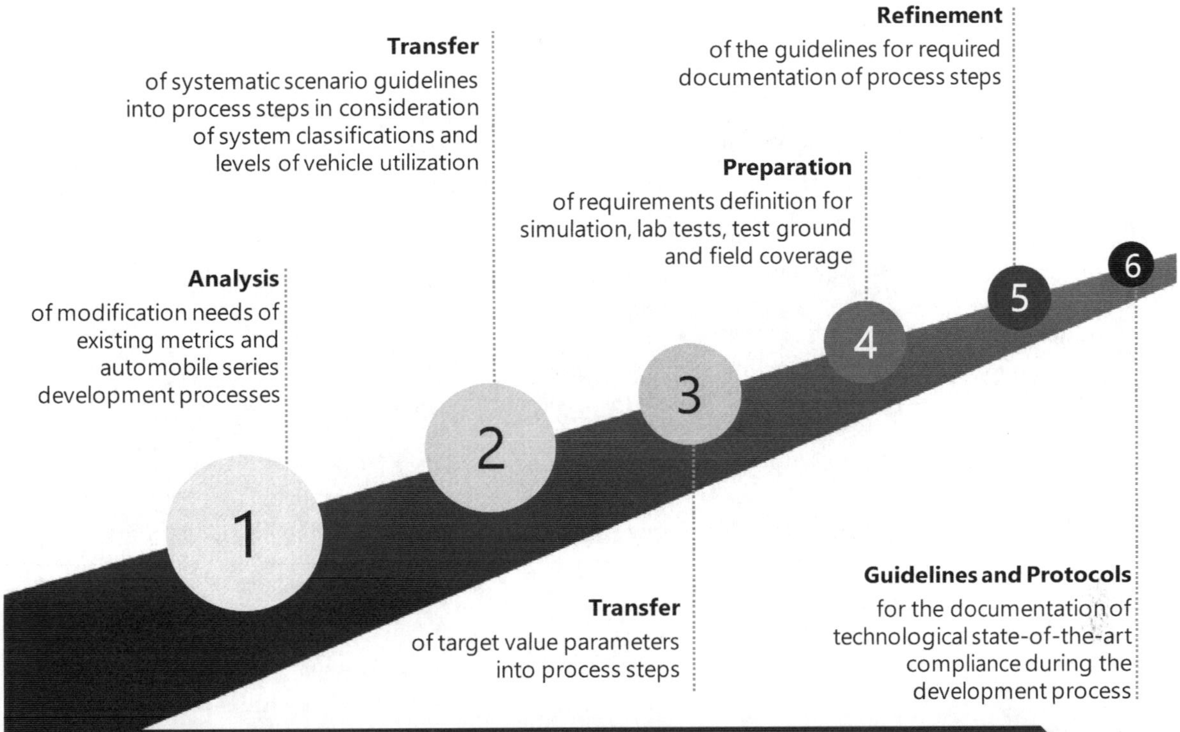

Central Issues of the PEGASUS Project

What level of performance is expected of an automated vehicle?
How can we verify that it achieves the desired performance consistently?

Scenario Analysis & Quality Measures | Implementation Process | Testing | Reflection of Results & Embedding

Testing

How can completeness of relevant test runs be ensured?

What do the criteria and measures for these test runs look like?

What can be tested in labs or in simulation? What must be tested on test grounds, what must be tested on the road?

PEGASUS

Testing

Generation of scenarios: levels of abstraction

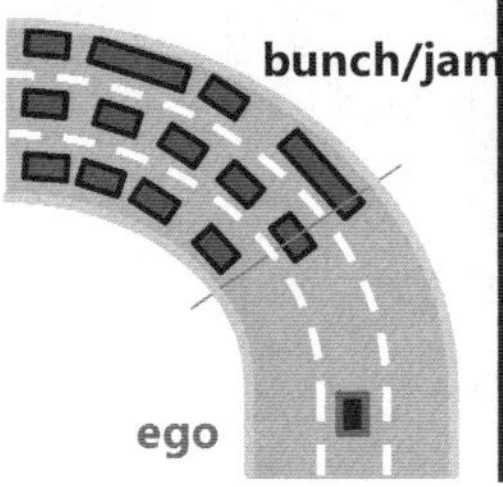

Functional scenarios	Logic scenarios	Concrete scenarios
Basis road: highway in bend	Basis road: number of lanes [2..4]; curve radius [0,6..0,9] kph	Basis road: number of lanes 3; curve radius 0,7 km
Stationary objects: -	Stationary objects: -	Stationary objects: -
Movable objects: ego, jam; interaction: ego approaches end of jam	Movable objects: End of jam position[10..200] m; jam speed [0..30] kph; ego distance [50..300] m; ego speed [80..130] kph	Movable objects: end of jam position 40 m; jam speed 30 kph; ego distance 200 m; ego speed 100 kph
Environment: summer, rain	Environment: temperature [10..40] °C; droplet size [20..100] µm; rain amount [0,1..10] mm/h	Environment: temperature 20 °C; droplet size 30 µm; rain amount 2 mm/h

number of scenarios

level of abstraction

Source: Lemmer VDA 2017

PEGASUS

Testing

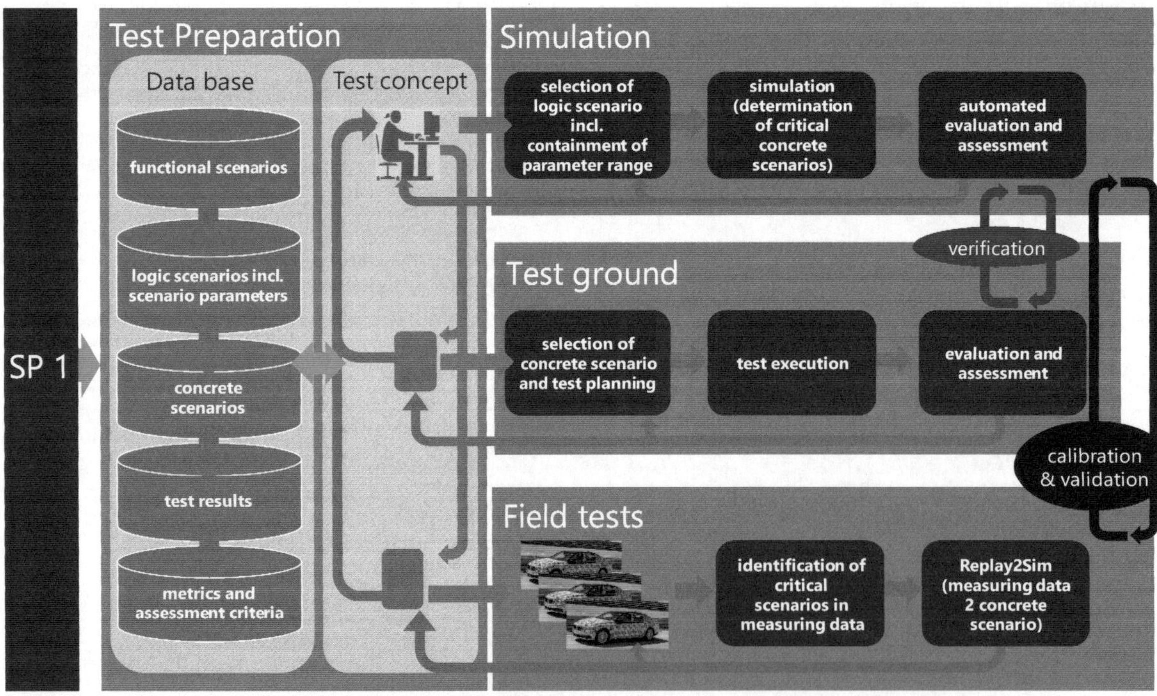

Central Issues of the PEGASUS Project

PEGASUS

What level of performance is expected of an automated vehicle?
How can we verify that it achieves the desired performance consistently?

Scenario Analysis & Quality Measures | Implementation Process | Testing | Reflection of Results & Embedding

Is the concept sustainable?

How does the process of embedding work?

Reflection of Results & Embedding

PEGASUS

Statement
about the distribution ratio between the applied test methods (from simulation to test ground to field test)

Proof of Concept
trough verification (1), assessment (2) and statement (3)

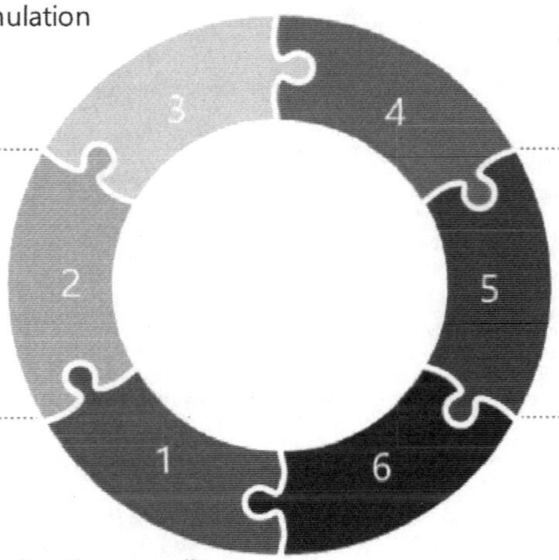

Assessment,
whether the test goal can be achieved with the utilized processes and methods in PEGASUS

Assistance
with embedding of acquired results with our project partners

Verification
of methods to identify relevant situations, quality and criticality measures for the assurance of HAD features

Lessons learned
regarding the implementation of the results in existing corporate structures

PEGASUS Goals beyond Research

- PEGASUS is a national project implementation for fast progress in automated driving.
- Embedding of findings in the industry.
- Distribution and pioneering of a standardization.

➔ All essential project results are freely accessible.

- Collaboration with other consortia is highly appreciated.
- We need a worldwide common understanding about how safety of automated driving has to be assured.
- Exchange with safety assurance experts worldwide at PEGASUS Symposium with interim presentation (9th of November 2017)

PEGASUS Symposium Nov. 9, 2017, Aachen

PEGASUS

PEGASUS Symposium - HOW SAFE IS SAFE ENOUGH?

09. November 2017

Institute for Automotive Engineerng (ika)
Aachen

ABOUT THE EVENT

HOW SAFE IS SAFE ENOUGH? HOW TO PROOVE THAT THE AUTOMATION IS GOOD ENOUGH?
17 scientific and industry partners are working in the joint research project PEGASUS to answer these questions, define a new state of the art for the test and release of highly automated driving functions and demonstrate the procedure hands-on using a sample application - autonomous Autobahn-driving.

With the **PEGASUS Symposium**, an international platform will be created, which allows an **open exchange for effective and uniform testing** as well as for securing highly-automated vehicles.

The PEGASUS Symposium will be accompanied by an **exhibition** which provides insights of the work in PEGASUS and presents the major results of the first half of the project.

AGENDA

Thursday, 09. November 2017

08:30	Registration
09:00	Welcome
09:15	Keynotes
11:15	General introduction to PEGASUS opening of the parallel exhibition
13:00	Automated driving initiatives and projects
15:30	PEGASUS key activities
16:30	Panel Discussion
17:30	Dinner speech
	Evening reception

register online at: www.pegasusprojekt.de

References:

Wachenfeld et al. 2016: Wachenfeld, W.; Junietz, P.; Winner H.; Themann, P.; Pütz, A.: Safety Assurance Based on an Objective Identification of Scenarios – One Approach of the PEGASUS – Project; Presentation TRB-AUVSI-Automated Vehicle Symposium, San Francisco, 2016-07-20

Lemmer, K., VDA, 2017: PEGASUS: Automatisiertes Fahren effektiv absichern (PEGASUS: effectively ensuring automated driving), Presentation slides, VDA Technial Congress 2017

Contact:
Prof. Dr. Hermann Winner
Institute of Automotive Engineering
Technische Universität Darmstadt

winner@fzd.tu-darmstadt.de

www.pegasusprojekt.de

PEGASUS

Supported by:

Federal Ministry for Economic Affairs and Energy

on the basis of a decision by the German Bundestag

Let's Move the Security Needle: Think Offensively!
Dr. Jonathan Petit
jpetit@onboardsecurity.com
OnBoardSecurity

UT Austin Researchers Successfully Spoof an $80 million Yacht at Sea
Low level vulnerabilities that led to high level exploits
Team of hackers take remote control of Tesla Model S from 12 miles away
ONBOARD SECURITY 7/17/2017
2

So… how come we haven't seen much changes since?

- "I'm busy getting it work. I'll deal with that later." *Anonymous AV Engineer*
- "I'm just a tiny piece of the puzzle. Who cares?" *Anonymous IMU Engineer*
- "These attacks don't scale so it isn't priority 1." *Anonymous Tier 1 Security Researcher*
- "We are working on it but don't publish about it." *Anonymous OEM Security Researcher*

ONBOARD SECURITY 7/17/2017 3

How do we move the needle?!

ONBOARD SECURITY 7/17/2017 4

AUTOMATED VEHICLES SYMPOSIUM 2017

USERS. VEHICLES. INFRASTRUCTURE.

Think Offensively, Share Proactively

We want to change how you perceive the world, so that you can see it through the eyes of a hacker. We ask that you then share with us any thoughts or ideas you might have. We'll compile the submissions and summarize them into a report to be presented at the end of AVS 2017.

Send your thoughts to: **AVS2017CS@protonmail.com**

Those who submit input will be automatically entered into a raffle for a $50 Amazon Gift Card

ASK YOURSELF

- **Can a normal function be weaponized?**
- **How easy is it to monetize a vulnerability?**
- **What are the dependancies of your work?**

Feel free to reach out to the following Cyber Security Working Group members on-site:

Edward Fok	Tim Leinmueller	Jonathan Petit
USDOT	Denso	OnBoard Security

Cybersecurity resources available here:
http://www.automatedvehiclessymposium.org/program/2017-speakers/jpetit

http://www.automatedvehiclessymposium.org/program/2017-speakers/jpetit

References to Known Automotive and Infrastructure Vulnerabilities

1. Modern Vehicle Vulnerabilities: http://illmatics.com/carhacking.html
2. Center for Automotive Security: http://www.autosec.org/publications.html
3. DEFCON Car Hacking Village: http://www.carhackingvillage.com/
4. Car Hacking Handbook: http://opengarages.org/handbook/
5. Vulnerabilities to Machine Learning
    1. Tailored input to machine vision learning set not detectable by human examiners caused the classifier to retu
    2. https://www.youtube.com/watch?v=JAGDpJFFM2A
    3. https://blog.openai.com/adversarial-example-research
6. Automated vehicle attacks over sensor inputs

1.GPS:

1. https://news.utexas.edu/2013/07/29/ut-austin-researchers-successfully-spoof-an-80-million-y
2. Countermeasure: https://utexas.influuent.utsystem.edu/en/publications/gps-spoofing-detection-v of-military-

2. RADAR, Ultrasonic: https://www.youtube.com/watch?v=orWqKWvIW_0

3. LIDAR, Camera:

1. https://www.youtube.com/watch?v=JAGDpJFFM2A
2. http://eprint.iacr.org/2017/613

4. TPMS:

1. http://www.youtube.com/watch?v=bKqlq2Y43Wg
2. http://www.youtube.com/watch?v=TDYoo7TGNcw

5. LTE: https://conference.hitb.org/hitbsecconf2013ams/materials/D1T2%20-%20Philippe%20Langlois%20-%20Hacking%20HLR%20HSS%20and%20MME%20Core%20Network%20Elements.pdf

7. Why is this a target – More on how hackers think?

1. What is the value of a hacked email
2. https://krebsonsecurity.com/2013/06/the-value-of-a-hacked-email-account/
3. What is the value of a hacked PC
4. https://krebsonsecurity.com/2012/10/the-scrap-value-of-a-hacked-pc-revisited/

8. Platoon centric attacks

1. Vehicular Platooning in an Adversarial Environment: Gerdes, Ryan; Virginia Tech
2. Attack Mitigation in Adversarial Platooning Using Detection-Based Sliding Mode Control; Gerdes, Ryan; Virginia Tech

Consortium/Standardization activities:

- Future of Automotive Security Technology Research (FASTR): www.fastr.org
- Securing Smart Cities (SSC): http://securingsmartcities.org
- I am the cavalry: https://www.iamthecavalry.org
- Defense Automotive Technologies Consortium (DATC): http://datc.saeitc.org
- Automotive Consortium for Embedded Security (ACES): aces.swri.org
- Open Automotive Alliance (OAA): https://www.openautoalliance.net
- UN Task Force on Cyber Security and OTA issues: https://wiki.unece.org/pages/viewpage.action?pageId=40829521
- SAE: J3061, J3101, Data Link Connector Vehicle Security Committee (https://www.sae.org/servlets/works/committeeHome.do?comtID=TEVDS20)
- TRB: Cybersecurity Subcommittee, Task Force on Data Privacy, Security and Protection Policy (https://www.mytrb.org/CommitteeDetails.as CMTID=4553)

25 breakout sessions = 25 good reasons to think offensively

- (#2+25) Human Factors
  - What is the impact of attack (e.g. billboard hacked) on user reaction?
  - How to re-engage user to mitigate ongoing attack?
- (#4) An AV Crashes: What Happens Next?
  - How to ensure chain of custody or detect tampering?
  - What data is needed to detect that crash was due to the car being hacked?
- (#6) Trucking Automation: Key Deployment Scenarios
  - What attacks are possible on automated trucks?
  - Should we consider vulnerabilities to automated trucks only or across the supply chain (higher scalability)?

ONBOARD SECURITY 7/17/2017

7

25 breakout sessions = 25 good reasons to think offensively

- (#9) Effects of Vehicle Automation on Energy-Usage and Emissions
  - How can an attacker affect energy-usage and emissions (e.g. forcing vehicles to drive longer, energy depletion attack)?
  - Will AV expose (more) the smart grid to potential attack?
- (#10) Data Sharing Models and Policy
  - Where/How to securely store the data?
  - Can we "watermark" the data to prevent unauthorized sharing?
- (#11) Artificial Intelligence
  - How Machine Learning technique (e.g. DNN) can be fooled?
  - How to address unintended consequences of emergent behavior and embed resilience?

ONBOARD SECURITY 7/17/2017

8

25 breakout sessions = 25 good reasons to think offensively

- More online!

http://www.automatedvehiclessymposium.org/program/2017-speakers/jpetit

Dr. Petit is the Senior Director of Research for OnBoard Security (formerly known as the Embedded Security Business Unit of Security Innovation). He is in charge of leading projects in security and privacy of automated and connected vehicles. He has conducted extensive research in detecting security vulnerabilities in automotive systems. He published the first work on potential cyber attacks on automated vehicles and has supported communications security design and cybersecurity analysis through OEM and NHTSA-sponsored projects. He received his PhD in 2011 from Paul Sabatier University, Toulouse, France.

The following Excel file (and pdf version) lists security-related questions for each breakout sessions in order to foster discussion and get your feedback. Please don't restrict yourself to these questions. Send us your questions / comments!

ONBOARD SECURITY 7/17/2017 Let's Move the Security Needle: Think Offensively! 9

Thank you and please give us your input!

AVS2017CS@PROTONMAIL.COM

You have a PhD in Computer Security?
OnBoard Security is hiring!

OnBoardSecurity

GTAI GERMANY TRADE & INVEST

Outline

1. **Executive Summary**
   a) The Industry in numbers
   b) The Premium Segment

2. **Market Trends**
   a) Overview of Maket Trends
   b) Disruptors in Automotive Industry
   c) Autonomous and Connected Car Technologies

3. **GTAI - Investor Support**
   a) Who we are
   b) What we do

GTAI GERMANY TRADE & INVEST

1. EXECUTIVE SUMMARY

Turnover in the Automotive Industry

Germany´s automotive industry turnovers exceeded 400 billion EUR

Turnover of the German Automotive Industry in Germany
(in billion EUR)

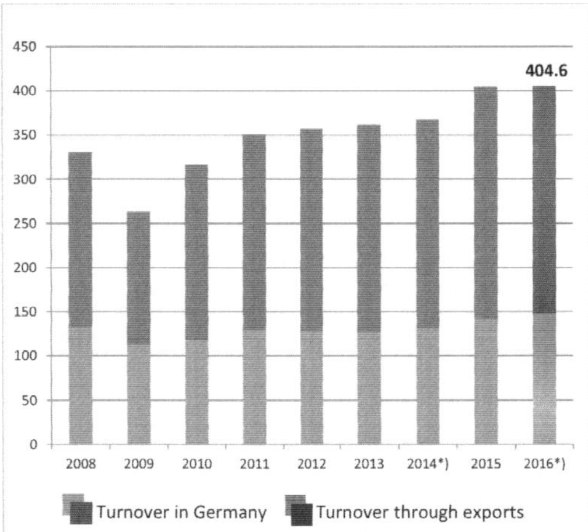

Total turnover of **EUR 404.6 billion**

High export share: EUR 256.3 billion generated in foreign markets.

79% of the industry's turnover (EUR 317 billion) is **generated by the OEMs**

Source: VDA (2017)

Germany´s Automotive Market Size

Germany is Europe´s largest automotive market

Passenger car production in Europe 2016
(in million units)

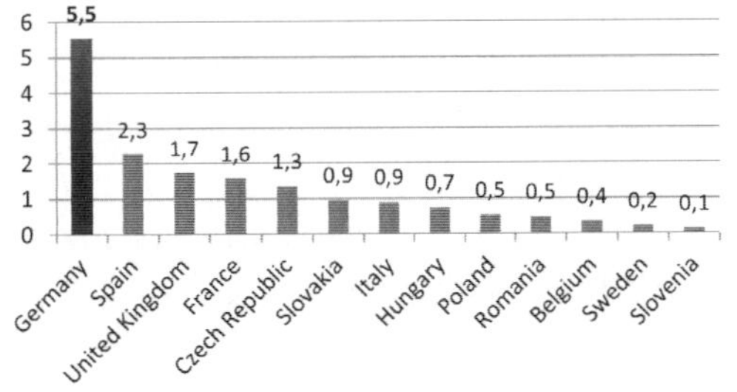

Increase of passenger car production numbers in Germany in 2016: **+0,2%**

Number one market by production and sales, accounting for over 33% of all passenger cars produced in Europe

Passenger car registrations in Europe 2016
(in million units)

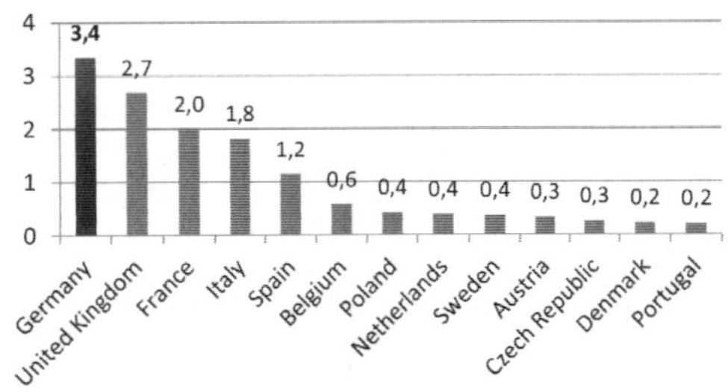

Increase of passenger car registration numbers in Germany in 2016: **+4,5%**

Number one market by car registrations, accounting for over 22% of all newly registered passenger cars in Europe

Sources: ACEA (2017); OICA (2017)

Industry R&D Investments

German automotive industry is the leading innovator

Internal R&D investments in Germany within the industrial sector
(2015, in billion EUR)

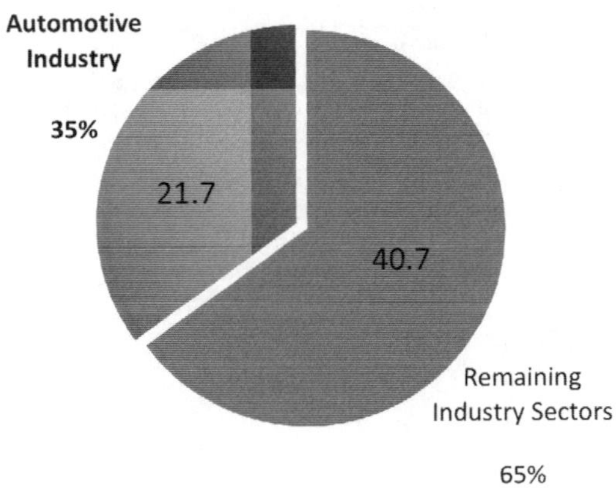

R&D expenditures of the German automotive sector account for **35% of industrial R&D spending** & **increased by 10%**

Germany is world leader in auto industry patents: German OEMs account for more than **1/3rd of the world´s auto R&D**

110,000 highly skilled engineers work on automotive-related R&D-topics in Germany (+10% increase to 2015)

Sources: GTAI Research (2016), VDA (2017), Stifterverband (2016)

The Premium Segment

Germany: World´s premium car production and innovation hub

Premium car production by country 2016

PORSCHE Audi

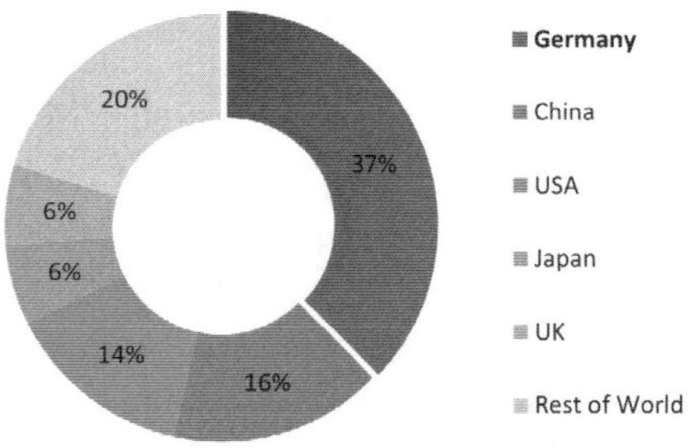

37% of all produced premium vehicles are made in Germany

56% of all produced premium vehicles are manufactured in Europe

68% of the premium car production is manufactured by German OEM

The country's **strong industrial base**, its **value chain density**, its **R&D power** and particular **strength in the automotive premium business, enables investors** to **develop cutting edge automotive technologies** for today's automotive needs. [...]"

Sources: GTAI Research (2017); MarkLines database (2017)

GTAI GERMANY TRADE & INVEST

Mega trends in the automotive industry

Efficient and connected cars

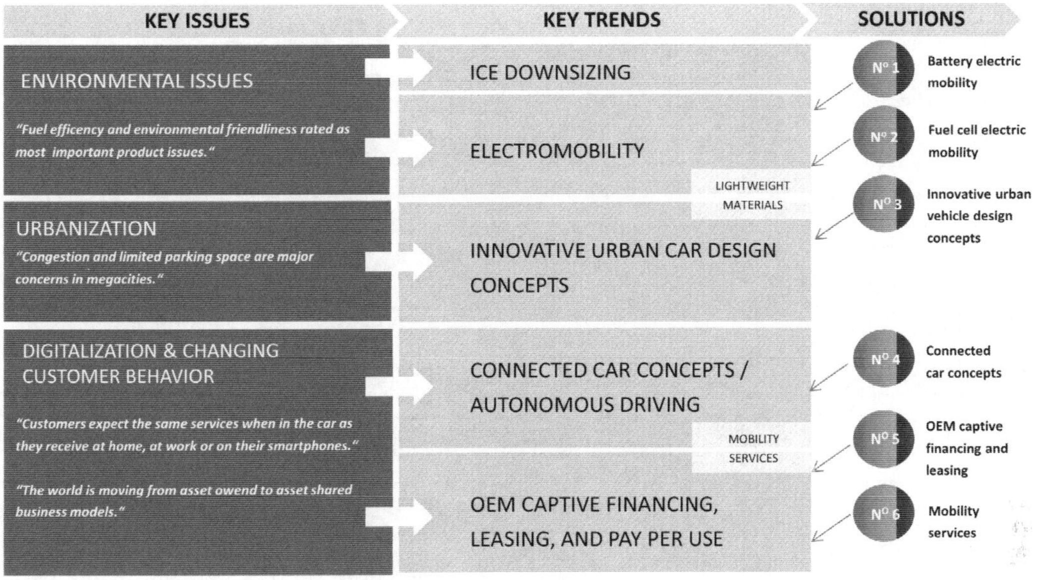

Source: KPMG's Global Auto Executive Survey (2012); GTAI Research (2016)

Miles will Matter in the Future

New definition of the business model

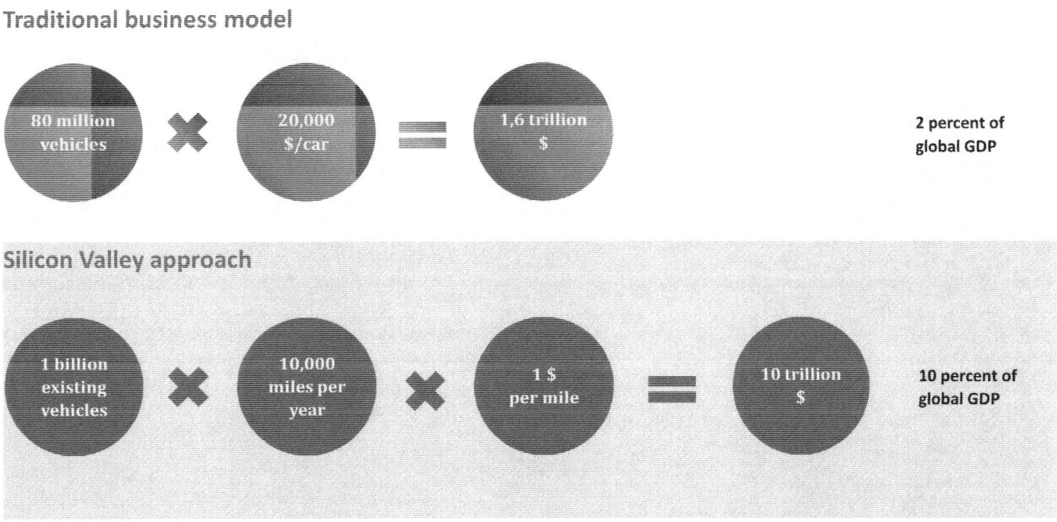

Sources: A . Jonas, Morgan Stanley (2016); GTAI Research (2016)

National Strategy – Automated & Connected Drive

National strategy with five central action fields

Roundtable Automated Drive

- **Foundation:** Set up by the Federal Ministry of Transport and Digital Infrastructure in 2013
- **Involved Parties:** Experts from the fields of politics, R&D, insurance industry, vehicle manufacturer and suppliers
- **Main Target:** Development of strategic implications to answer the numerous individual legal and economic aspects related to automated driving

Strategy development based on work results *Ongoing advancement & exchange*

National Strategy Automated and Connected Drive

Federal Ministry of Transport and Digital Infrastructure

Goal	**Remain Leading Technology Provider**	**Become Lead Market**	**Enable automated and connected drive**		
Potential	Increase of transport efficiency & quality	Increase of transport safety	Reduction of mobility-related emissions	Reinforcement of Germany as business location	
Action Fields	**Infrastructure**	**Law**	**Innovation**	**Connectivity**	**IT-Security**

Source: Federal Ministry of Transport and Digital Infrastructure ; GTAI Research

Automated Drive Landscape

Digital Test fields and national automotive R&D & industry

Three different test fields enabled for autonomous and automated driving

Initiator:

- Federal Ministry of Transportation and Digital Infrastructure; State Ministries Baden-Württemberg and Lower-Saxony

Extract Test fields:

- **Bavaria:** Automated driving on the Autobahn A9 – Real time measurement of dynamic data; for more safety and better traffic flow
- **Baden-Württemberg:** Autonomous and automated action field Karlsruhe; urban and rural areas will both be tested
- **Lower-Saxony**: Approximately 270 km length between Hanover, Brunswick, and Salzgitter by 2018

Target:

- Testing, evaluating & improving mobility 4.0 technologies

Source: Federal Ministry of Transport and Digital Infrastructure; Federal Ministry of Economic Affairs and Energy; GTAI Research

Digital Testfield Autobahn (A9)

Germany: Test Area in Bavaria (Federal level)

Testing:

- Enabling testing for self driving, highly automated vehichles (level 3)
- Newly developed HD Map
- 5G car-to-car real-time communication standard
- Extraction of real-time data

New Traffic Sign:

- On Autobahn A9 between Ingolstadt and Munich
- Not ment for drivers
- Serves as orientation landmark for automated cars

Incentives for Investors Guidline:

- Focus lies on innovative infrastructure elements to enhance testfields
- No deadlines for submission
- However, projects must be completed no later that end June 2019

Sources: Federal Ministry of Transport and Digital Infrastructure

Test Area For Automated Driving Karlsruhe

Germany: Test Area in Baden-Württemberg (State level)

- Baden-Württemberg is strengthening its position in transportation and the automotive industry towards autonomous and connected driving.

Operational Fields

Vehicles & Infrastructure	Business Models & Mobility Concepts	Legal Aspects & Validation	Social Acceptance

Application Areas

Automated Commercial Vehicles: public shuttle & bus transport, logistics & supplier operation	Automated Passenger Cars: car sharing, communting traffic, individual transport, micro mobility	Services: mobility apps and services

Sources: e-Mobil BW

Test Track for Autonomous Driving

Germany: Test Area in Lower-Saxony (State level)

- Niedersachsen's automotive sector leads Europe on automotive digitalization for autonomous driving
- On the Lower-Saxony test track, new technologies are being tested in real driving conditions. Unique within Europe.
    - Open research platform
    - 280 Km test track (highways, freeways, and urban traffic conditions)
- The project is led by the State of Lower-Saxony and the transport research department of the German Aerospace Center (DLR)

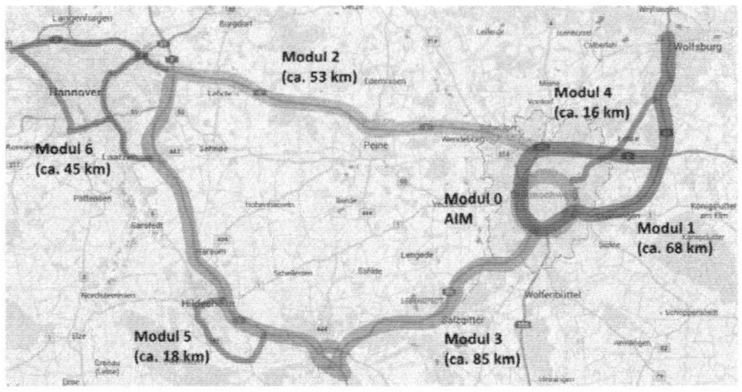

Project partners:

Niedersachsen DLR

Sources: State of Lower-Saxony, DLR

Automated Drive Roadmap

Unified agreement about the process towards the autonomous car

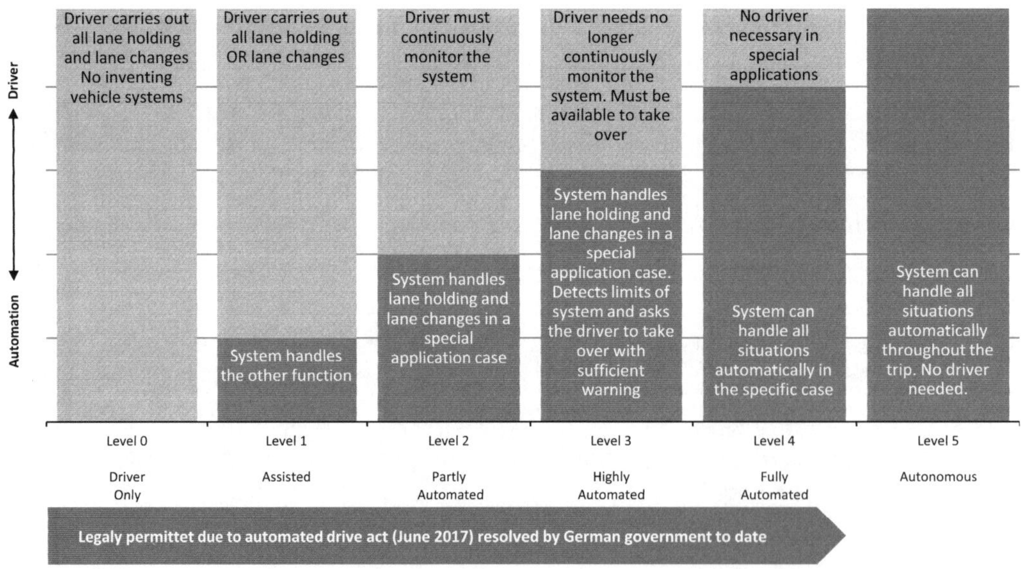

Source: Federal Highway Research Institute; German Association of the Automotive Industry (VDA) 2015

Autonomous drive in Germany

German OEMs and suppliers are dominating autonomous drive innovations

Amount of patents related to autonomous drive technologies
(By different company groups and Top-10 patenting companies; 2010-2016)

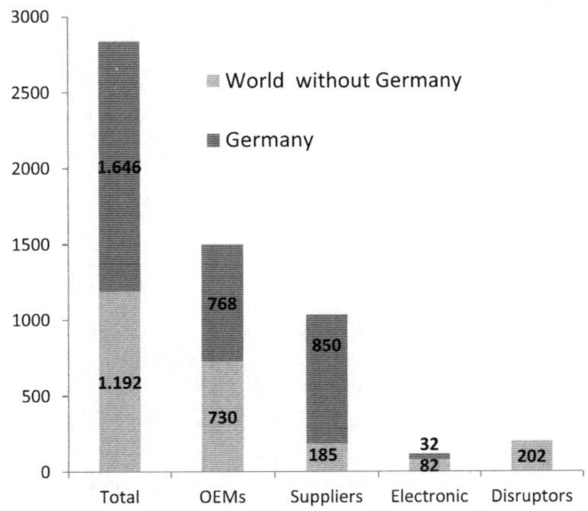

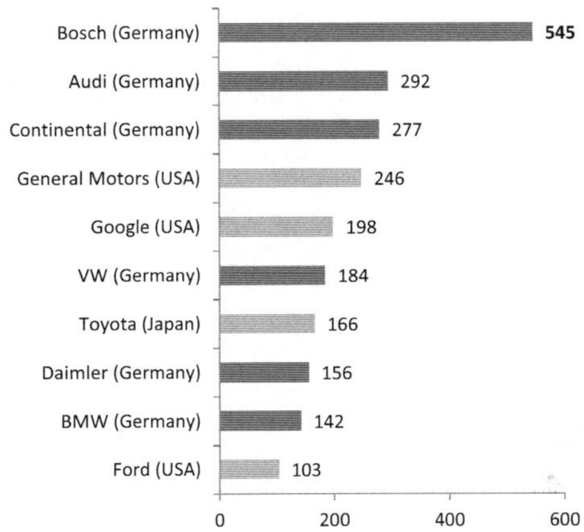

**Disruptors, e.g. Google, Apple*
Sources: PATENTSCOPE (2016); IW, Köln (2016)

GTAI GERMANY TRADE & INVEST

4. GERMANY TRADE AND INVEST - SUPPORT

Who we are

 www.gtai.com 19

 www.gtai.com 20

Advantages of the partner network

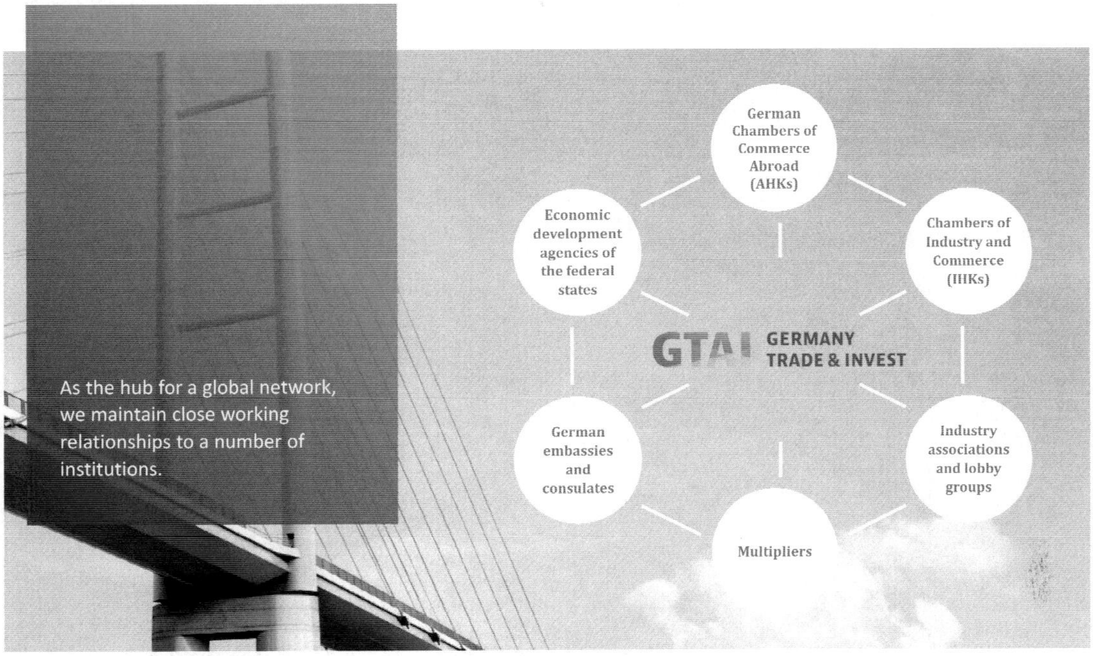

About GTAI Support

Two headquarters in Germany – 50 locations worldwide

Germany Trade & Invest serves as a one-stop consultant to the investing company

Investor

GTAI GERMANY TRADE & INVEST

Economic Development Agencies of the Federal States

ZAB, WFB, HESSEN TRADE & INVEST, M INVEST, gwSaar, HWF, invest in bavaria, Niedersachsen, Sachsen-Anhalt, bw-i, WTSH, LEG, Berlin Partner, NRW INVEST, Wirtschaftsförderung Sachsen, ISB

Investment consultancy services

- Discussion on the market/industry and competition in Germany and Europe
- Information on **tax and legal issues** in Germany
- Information on different **registration or licensing requirements**
- Information on the **financing of the project** (including incentives)
- Discussion of **location proposals/site evaluation incl. site visits**
- Joint project management with regional economic development agencies

Velodyne LiDAR, Inc.

Automotive sensors and real time 3D data

Company profile (prior to investment)

- **About:** Velodyne was founded in 1983 as an audio company specializing in low-frequency sound and subwoofer technology. Starting in 2007, Velodyne expanded into LiDAR technology
- **Industry:** Since 2007, Velodyne LiDAR has focused on applications of LiDAR technology for use in autonomous vehicles, vehicle safety systems, 3D mobile mapping, 3D aerial mapping and security.
- **Products:** Mainly LiDAR sensors. They are the most advanced in the market.

Project data

- Initial Investment: Approximately up to 100.000 Euro
- Location: Velodyne Europe GmbH in Rüsselsheim
- FTE's: 20
- First contact with GTAI: January 2016
- Customer service center

Milestones

Date	No.	Milestone
January 2016	1	First contact with Velodyne at CES 2016, NV., USA
January 2016	2	General information regarding Germany as a business location ,and automotive hub
February 2016	3	Meeting in Berlin with Erich Smidt, GF Dt.
March 2016	4	Mediate to Hessen state agency
March 2016	5	Transfer project to Hessen state agency

Impressions

Velodyne LiDAR®

Thank you for your attention

Feel free to visit us at booth #307:

GTAI GERMANY TRADE & INVEST

Mr. Stefan Di Bitonto
Senior Manager Investor Consulting
Transportation Technologies
T. +49 30 200 099 401
stefan.dibitonto@gtai.com

Saving Lives Now With Automation

Kelly Nantel
V.P., Communication and Advocacy
National Safety Council
Kelly.Nantel@nsc.org

Alex Epstein
Sr. Director, Digital Strategy and Content
National Safety Council
Alex.Epstein@nsc.org

Eliminating Preventable Deaths

The National Safety Council eliminates preventable deaths at work, in homes and communities, and on the road through leadership, research, education and advocacy

2

Eliminating Preventable Deaths

It's All About Improving Safety

An estimated 40,000 deaths and rising

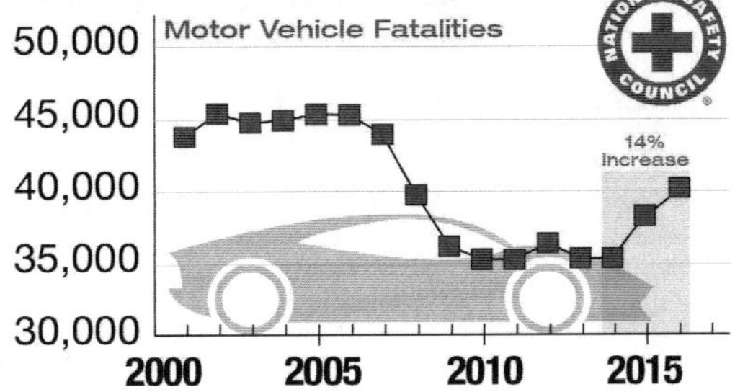

Eliminating Preventable Deaths

3

Key Transportation Initiatives

Road To Zero

National Safety Council, National Highway Traffic Safety Administration, Federal Highway Administration and Federal Motor Carrier Safety Administration, announced the Road to Zero initiative. The aim is to eliminate traffic fatalities within the next 30 years.

NSC Defensive Driver Courses

Train over one and a quarter million drivers each year

Teen – GDL

Continued work in passing stronger GDL laws – and parents of new teen driver campaign – DriveitHOME

Child Passenger Safety Seats / Hot Cars

Manage National Child Passenger Safety Board – support other advocates

Eliminating Preventable Deaths

4

Advanced Vehicle Safety Technology

Chair the ANSI (American National Standards Institute) Z15.3 autonomous vehicle subcommittee

Goal of developing best practices for fleet operations.

DOT Advisory Committee on Automation

Check to Protect

MyCarDoesWhat

National Driver Education Project on Advanced Vehicle Safety Technology

How Automation Can Help

- NHTSA assigned the critical reason for crashes (the last event in the crash causal chain) to be the driver in 94% of crashes investigated.
- When we consider the top three factors in crashes: alcohol, speed and distraction – fully automated vehicles that are not drunk, reckless or distracted have the potential to impact preventable deaths in an unprecedented way.

Source: USDot -NHTSA – Traffic Safety Facts – February 2015, Critical Reasons for crashes Investigated in the National Motor Vehicle Crash Causation Survey

Why Education Is Necessary

- Some ADAS Features Appear to Be Working
- Mixed Driver Understanding of How Features Work and How To Use Them
- Help Introduce Higher Levels of Automation to More Drivers

NSC Has Been Running Public Education Campaigns On Driver Safety For Decades

MyCarDoesWhat™.org

Know More. Drive Safer.

The University of Iowa

Eliminating Preventable Deaths

9

What Does This Icon Represent?

Eliminating Preventable Deaths

10

What Does This Icon Represent?

TPMS

Eliminating Preventable Deaths

11

What Does This Icon Represent?

Eliminating Preventable Deaths

12

What Does This Icon Represent?

13

Eliminating Preventable Deaths

What Does This Icon Represent?

Drowsiness Alert

14

Eliminating Preventable Deaths

What Does This Icon Represent?

Eliminating Preventable Deaths
15

What Does This Icon Represent?

Automatic Emergency Braking

Eliminating Preventable Deaths
16

Why This Initiative?

Why This Initiative?

© 2017 National Safety Council

Eliminating Preventable Deaths

ADAS Systems are the Building Blocks of Automation

Eliminating Preventable Deaths

Website

Eliminating Preventable Deaths

Easy Explanations

Facebook Games

App – Dashboard Blitz

Eliminating Preventable Deaths

What is MyCarDoesWhat?

- MyCarDoesWhat is the first of its kind – evidence based and independent
- Vehicle Agnostic
- Partnership between National Safety Council and University of Iowa
- More than 6 billion exposures – U.S. population 18+

Eliminating Preventable Deaths

Campaign Challenges

- Safety Features Have Different Brand Names
- Safety Features Have Different Capabilities across Manufacturers, Trim Levels and Time
- Safety Feature Limitations May Not Be Intuitive or Obvious
- Warning or Icon standardization issues

What You Can Do

- Suggest vehicles with "5-Star" ratings and advanced safety features
- Advocate for more collaboration amongst OEMs for more voluntary agreements on ADAS features
- Advocate for anonymized data sharing to keep track of how the systems work

What You Can Do

- Share MyCarDoesWhat.org with your co-workers, drivers, staff, families – It is a trusted, credible, non - OEM branded source
- Tell us what you think!

Saving Lives Now With Automation

Kelly Nantel
V.P., Communication and Advocacy
National Safety Council
Kelly.Nantel@nsc.org

Alex Epstein
Sr. Director, Digital Strategy and Content
National Safety Council
Alex.Epstein@nsc.org

Safe And Reliable: Link Process Models, ISO26262 and Cyber Security

San Francisco, 07/12/2017

MAGNA

Agenda

MAGNA

- Challenge: Unexpected Behavior in Modern Vehicles
- Automotive SPICE
- ISO26262
- Cyber Security

Our Global Presence

MAGNA

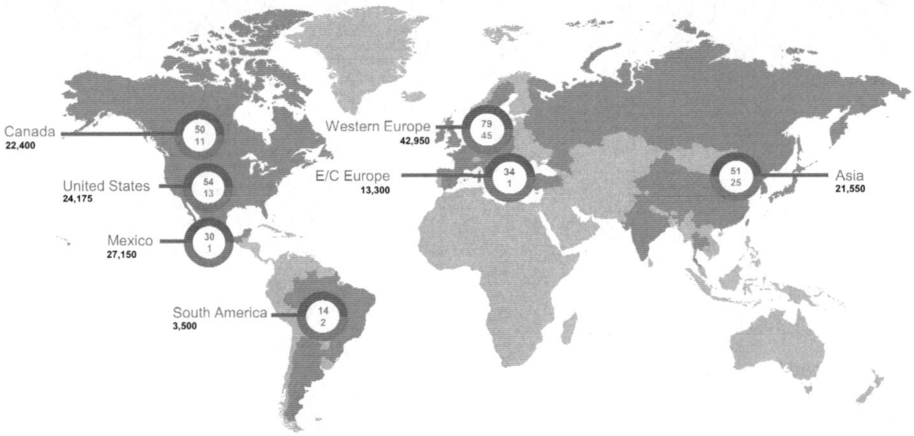

~155,000 People	29 Countries	● 312 ● 98	$32.1 Billion (2015 Sales)	● Manufacturing / Assembly ● Engineering / Product Development / Sales ● Number of Employees

@Q3 2016

Agenda

MAGNA

- Challenge: Unexpected Behavior in Modern Vehicles
- Automotive SPICE
- ISO26262
- Cyber Security

2017 Disclosure or duplication without consent is prohibited

Unexpected Behavior

MAGNA

Electronic systems can malfunction in the following ways:

a) Component HW failure
b) SW coding error
c) Requirement not implemented
d) Requirement misunderstood
e) Malicious external intervention

Prevention:

1. Define functional goals
2. Define safety goals
3. Protect from anomalies

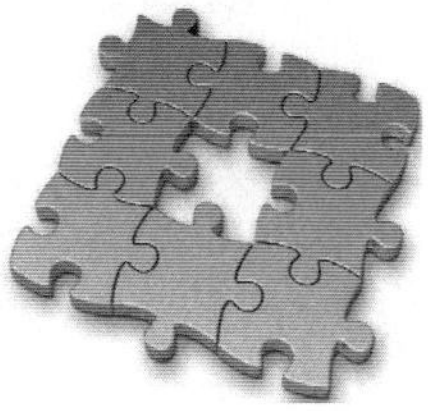

Design Hierarchies

MAGNA

A process model (Automotive SPICE or CMMI) is the foundation for ISO26262

ISO26262 ensures functional safety

Cyber security protects the system environment
- SAE J3061 Surface Vehicle Recommended Practice
- ISO 27001 and 27002 (Specify information security management and security management practice)
- ISO 15408 Common Criteria for Information Technology Security Evaluation

This presentation will only be able to highlight the key aspects of these three challenging areas.

Agenda

MAGNA

- Challenge: Unexpected Behavior in Modern Vehicles
- Automotive SPICE

Process Models

MAGNA

V-Model Approach
- From Generic to Specific
- Ensure requirements for specific step are completely understood
- Channel into disciplines

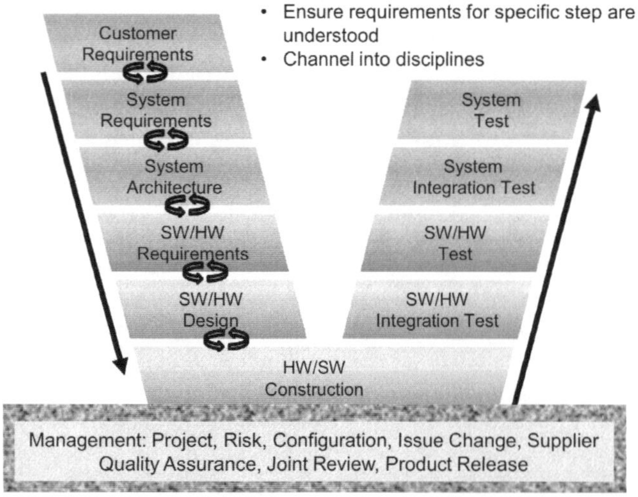

State of the art for Automotive:
- Automotive SPICE (ISO/IEC 15504, Software Process Improvement and Capability dEtermination)
- CMMI (Capability Maturity Model Integration)

Waterfall or Agile Methods may be used.

Define Requirements, System Elements And Interfaces

MAGNA

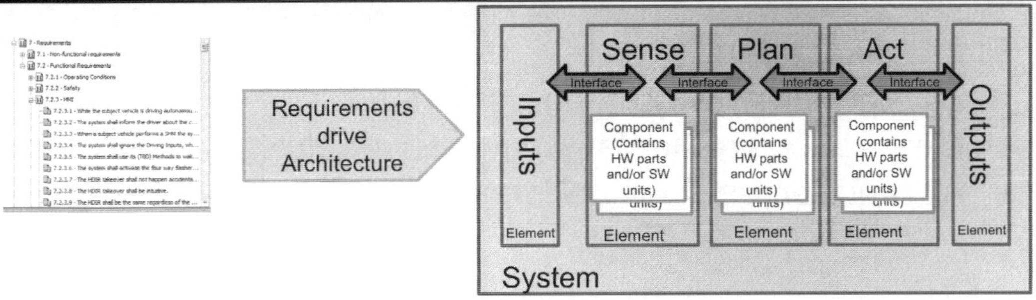

After decomposing the requirements into system requirements one can design the architecture ensuring that each requirement is represented by an architecture element. The interfaces between the elements need to be understood and defined.

This ensures:
- Requirements are understood
- Architecture represents the requirements

7/5/2017 Disclosure or duplication without consent is prohibited 9

Traceability

MAGNA

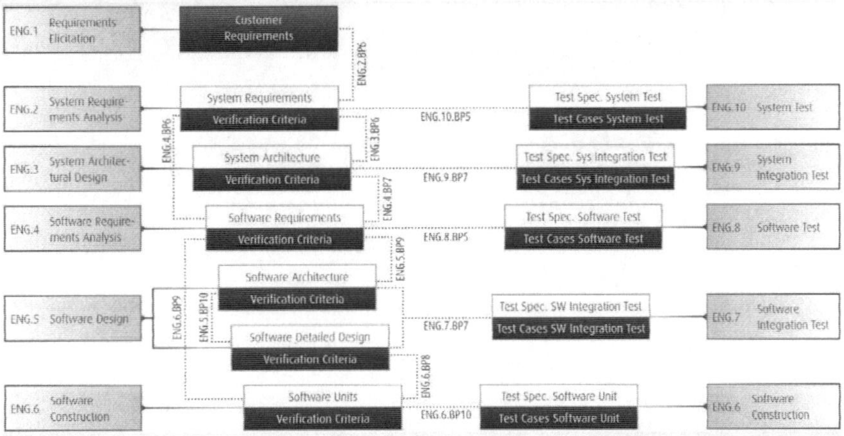

Identify non-implemented requirements through traceability
Ensure that all requirements were implemented and verified

7/5/2017 Disclosure or duplication without consent is prohibited 10

First Stage Complete

MAGNA

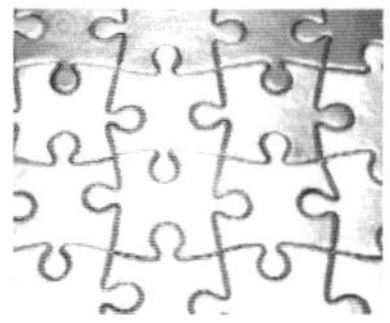

First achievement:

- All requirements are understood
- Missing requirements have been identified
- All requirements will be implemented

To Do:

- **Component HW failure**
- **SW coding error**
- **Malicious external intervention**

Agenda

MAGNA

- Challenge: Unexpected Behavior in Modern Vehicles
- Automotive SPICE
- ISO26262
- ...y

ISO26262: Goal

MAGNA

Goal: Avoid unreasonable risks!

Describe safety-related items:

1. Environmental dependencies
2. Interaction with the environment
3. Define safety life cycle activities
4. Identify hazards related to the item
5. Analyze hazards
6. Develop safety goals to prevent or mitigate the hazards
7. Decompose safety goals into functional safety requirements and then technical safety requirements

ISO26262: Classify The Risk

MAGNA

- Severity, exposure, controllability define Automotive Safety Integrity Level (ASIL A- D)
- ASIL represents a risk-based classification of a safety goal as well as the validation and confirmation measures required to ensure the goal is reached
  - Severity S: From S0 (no injuries) to S3 (survival uncertain / fatal)
  - Exposure E: From E0 (incredibly unlikely) to E4 (event could happen under most normal operating conditions)
  - Control C: From C0 (controllable in general) to C3 (difficult to control or uncontrollable)

S	E	C1	C2	C3
S1	E1	QM	QM	QM
	E2	QM	QM	QM
	E3	QM	QM	ASIL A
	E4	QM	ASIL A	ASIL B
S2	E1	QM	QM	QM
	E2	QM	QM	ASIL A
	E3	QM	ASIL A	ASIL B
	E4	ASIL A	ASIL B	ASIL C
S3	E1	QM	QM	ASIL A
	E2	QM	ASIL A	ASIL B
	E3	ASIL A	ASIL B	ASIL C
	E4	ASIL B	ASIL C	ASIL D

Source ISO 26262

- ➢ Table shows the combinations of S / E / C to determine the ASIL level
- ➢ Classification (QM - D) determines what measures have to be taken to mitigate the risk

ISO26262 Hazard & Risk Analysis

MAGNA

- Provides safety goals with ASIL classification
- Safe States (Should include examples)
- Defines fault tolerance time

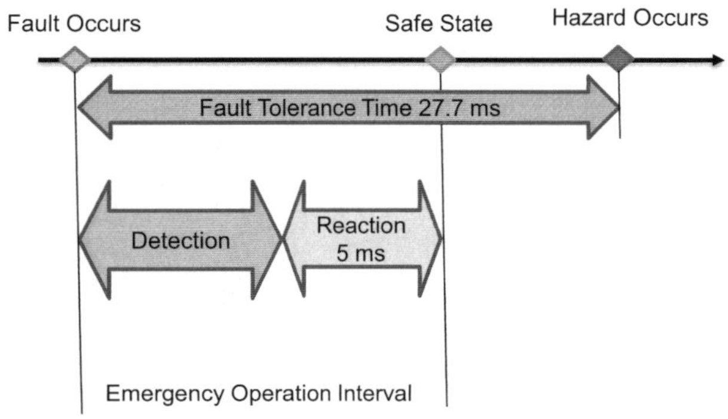

Example:
The ABS controller relies on a CAN network to transport the brake command. The system needs to react safely if the CAN network fails.
Safe State: Fallback on a redundant CAN

When one CAN network fails the backup CAN network shall be available for data transport within 5 milliseconds (reaction time).
Safety goal: The vehicle shall travel a maximum of 1 m at 130 km/h with no control (Fault tolerance time: 27.7 milliseconds)
Detection time < 22.7 ms (27.7 ms – 5 ms)

Second Stage Complete

MAGNA

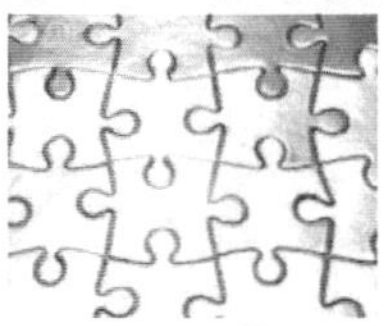

Second Achievement:

- **Component HW failure risk assessed & safety goals defined**
- **SW coding error risk assessed & safety goals defined**
- **Groundwork laid for detailed HW/SW design**

To Do:

- **Prevent malicious external intervention**

Agenda

MAGNA

- Challenge: Unexpected Behavior in Modern Vehicles
- Automotive SPICE
- ISO26262
- Cyber Security

Malicious Intervention

MAGNA

Types of malicious intervention (Ordered by severity)

1. Control the vehicle against the drivers wishes
   - Can cause injury / death
2. Theft (steal PIN, Credit Card information…)
   - Potential of severe economic damage to an individual (identity theft etc.)
3. Reduce features (Denial of Service)
   - User cannot use the feature that he / she paid for when purchasing the system

Prevent Vehicle Control

MAGNA

Functional Safety provides a layer of protection against unsafe behavior:

Safety Goals:
- Tolerate falsified messages
- Tolerate missing messages
- Tolerate wrong SW version on one ECU

BUT:
- Hackers could modify the whole system (source of vehicle speed signals, steering input from both steering wheel and lane keep assist and the power steering entity)
- The system would then enter an unsafe state controlled solely by the attacker

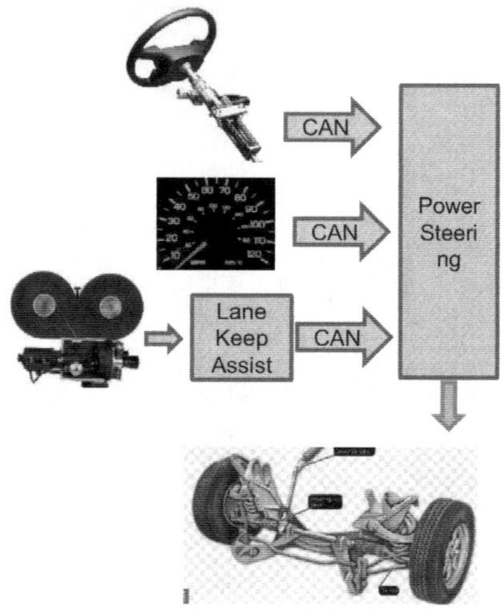

J3061 Gates

MAGNA

- Cybersecurity plan
- Feature Definition
- Threat Analysis and Risk Assessment
- Functional Cybersecurity Concept
- Functional Cybersecurity Requirements
- Initial Cybersecurity Assessment

From: J3161

All activities complete
Documents verified
Items validated
Cybersecurity case (Final Refined Cybersecurity Assessment)

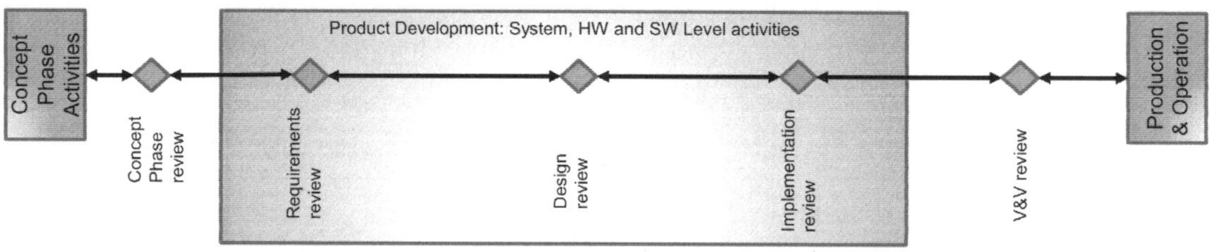

- Identify Requirements (Technical, Cybersecurity)
- Allocate Requirements to HW/SW
- Refined Cybersecurity Assessment

- Analysis Activities to Develop Design
- System Design
- Refined Cybersecurity Assessment

- Analysis Activities to Develop Implementation
- HW/SW Implementation
- Refined Cybersecurity Assessment

Layered Security Strategy

MAGNA

- Vehicle Level Security
- Module Level Security
- Board Level Security
- Chip Level Security

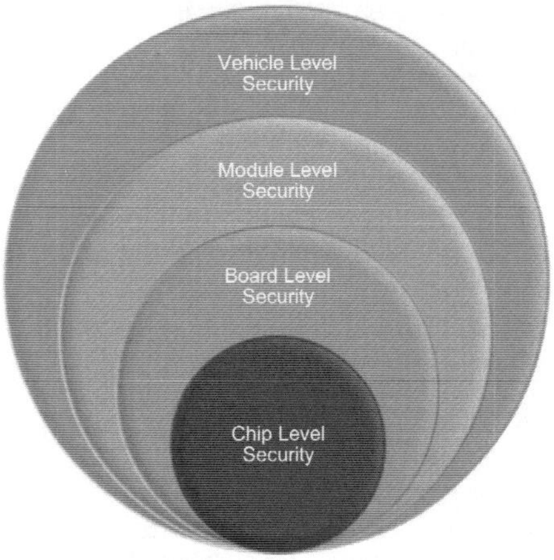

Security Gateway

MAGNA

- Firewall
  - Data Dictionary Enforcement
    - Timing check
    - Whitelist check
    - Message length check
  - Diagnostics Access Filter
    - Check service per role
  - Content Anomaly Detection
    - Blacklist check
- Intrusion Detection
  - Detect anomalies missed by the firewall
    - Multiple-message with irregular rate pattern
    - Multiple-message counter out-of-sequence
    - Message or diagnostic service inconsistent with vehicle operating state
    - Statistical message range violation
    - Statistical message value change rate violation
  - Collect new exploitations or vulnerabilities
- Denial of Service (DoS) Protection
    - Message ID counter (unique)
    - Several parameters calibrate-able

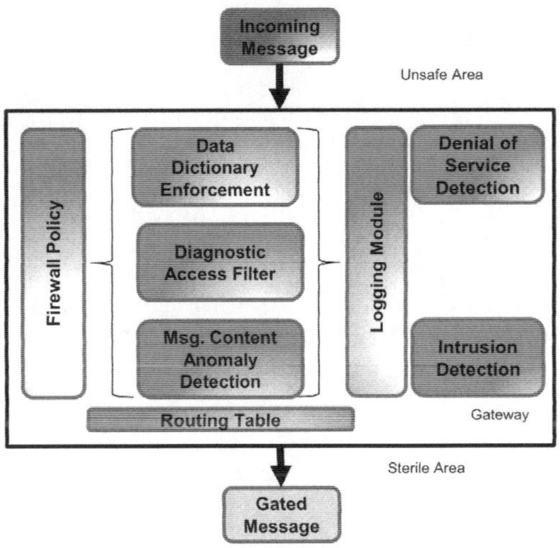

Module / Board / Chip Level Security

MAGNA

- **Secure Flash Programming**
  - Signature verification on the update file before execution
- **Secure Boot**
  - Parallel Boot – Boot ROM and App ROM check at the same time
- **Secure Key Store**
  - Keys are stored within HSM ROM
  - For usage, keys are only accessible by dedicated peripherals
  - For updates/overwrites, different keys need different authorizations
- **Secure Debug**
- **Secure Logging**
  - Access control to log
  - Each log entry is with monotonic counter and MAC
- **Secure Erase**
  - Overwrite rather than flag
- **Cryptographic Accelerators – AES128 and RSA2048**
- **True Random Number Generator (TRNG), TrustZone capable**

Secure Boot

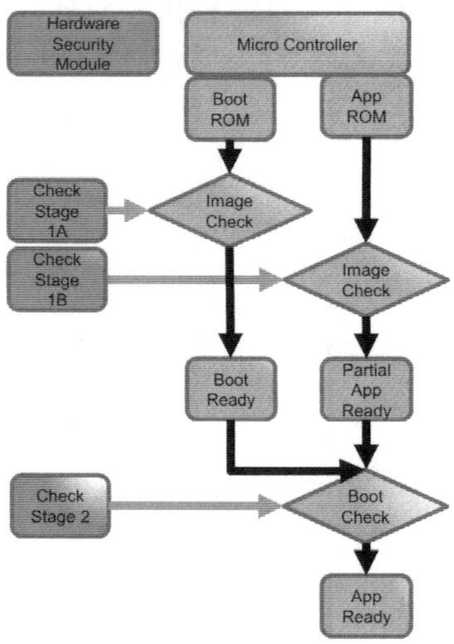

Secure Gateway Lifecycle Management

MAGNA

- Recovery Plan
- Lessons Learned
- Communications

- Asset Identification
- Threat Analysis
- Risk Assessment

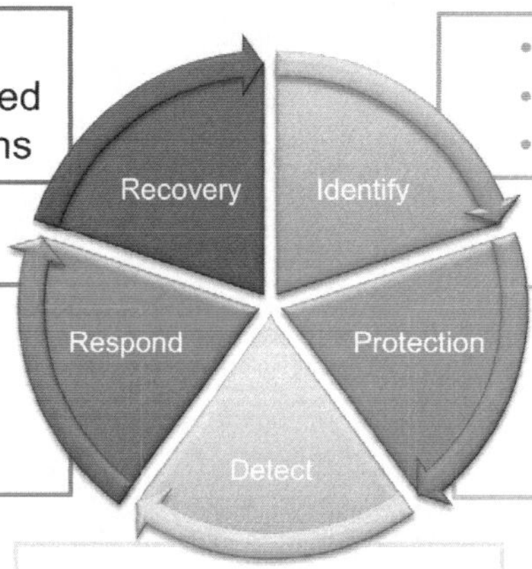

- IR Team
- Validation
- Containment Plan

- Access Control
- Data Security
- Awareness and Training

- Anomalies & Events
- Continuous Monitoring

Summary

MAGNA

Process
- Reduce development effort
- Increase confidence

A-Spice, CMMI

Functional Safety
- HW/SW malfunctions understood and mitigated

ISO 26262

Cyber Security:
- Intrusion detection and prevention

ISO 15408, ISO 27001, ISO 27002, J3061

Standards are the first building blocks…

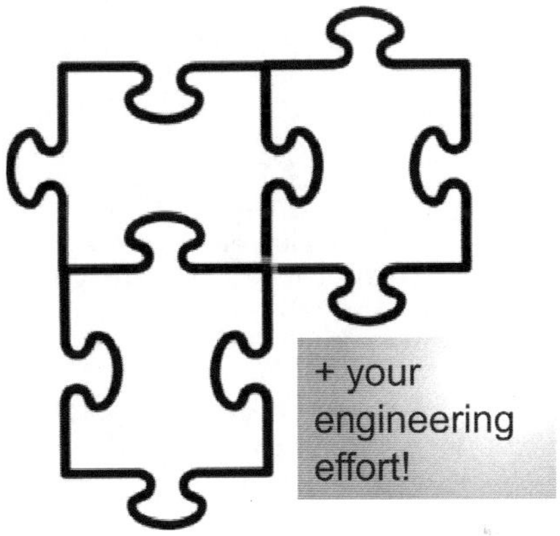

7/5/2017 Disclosure or duplication without consent is prohibited 25

Q&A

MAGNA

Industry Standards

CMMI (Capability Maturity Model Integration) is a process improvement approach across a specific set of processes, a project, a division or the entire organization.

Automotive SPICE is an industry-specific version of ISO/IEC 15504 or SPICE (Software Process Improvement and Capability dEtermination) which provides automotive industry specific process improvement and assessment methodologies.

HIS: “Hersteller Initiative Software”; association of German vehicle OEM; focused on the process areas in Automotive SPICE that impact the development process the most (15 out of 31 process areas)

The process model documents best practices from many organizations.
The process model describes **WHAT** to do, not **HOW** to do it.
Organizations need to determine how to apply the best practices of a process model to how the organization works in order to improve how work gets done.

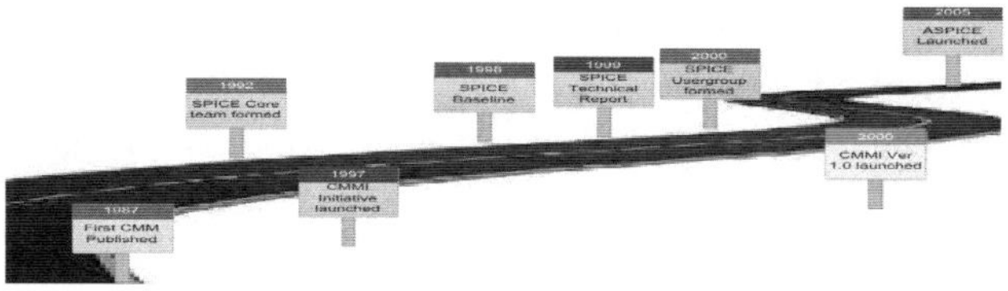

ISO26262

- Severity, exposure, controllability
- An Automotive Safety Integrity Level (ASIL) represents an automotive-specific risk-based classification of a safety goal as well as the validation and confirmation measures required by the standard to ensure accomplishment of that goal
  - At the beginning of the safety life cycle, hazard analysis and risk assessment is performed, resulting in assessment of ASIL to all identified hazardous events and safety goals.
  - Each hazardous event is classified according to the severity (S) of injuries it can be expected to cause:
  - Severity Classifications (S):S0 No InjuriesS1 Light to moderate injuriesS2 Severe to life-threatening (survival probable) injuriesS3 Life-threatening (survival uncertain) to fatal injuries
  - Risk Management recognizes that consideration of the severity of a possible injury is modified by how likely the injury is to happen; that is, for a given hazard, a hazardous event is considered a lower risk if it is less likely to happen. Within the hazard analysis and risk assessment process of this standard, the likelihood of an injurious hazard is further classified according to a combination of
  - exposure (E) (the relative expected frequency of the operational conditions in which the injury can possibly happen) and control (C) (the relative likelihood that the driver can act to prevent the injury).Exposure Classifications (E):E0 Incredibly unlikelyE1 Very low probability (injury could happen only in rare operating conditions)E2 Low probabilityE3 Medium probabilityE4 High probability (injury could happen under most operating conditions)Controllability Classifications (C):C0 Controllable in generalC1 Simply controllableC2 Normally controllable (most drivers could act to prevent injury)C3 Difficult to control or uncontrollable

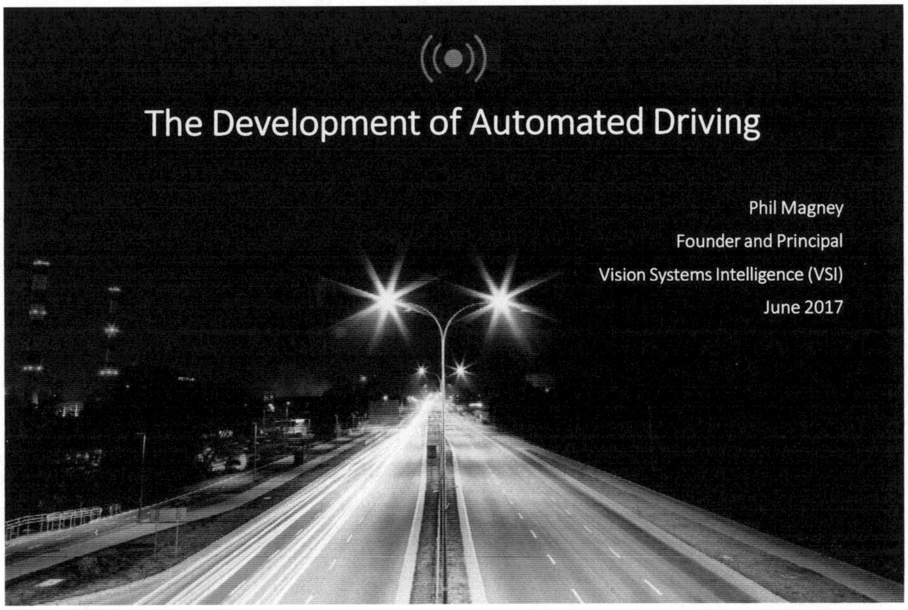

Who is VSI?

- VSI serves clients that are planning automated vehicle technologies.
- VSI research portals help R&D departments pre-qualify components, systems and development tools.
- VSI research portals help planning departments with technology roadmaps and detailed information about the direction of AV technologies.
- VSI Labs (part of VSI) examines leading components and tools through applied research using simulation tools as well as real vehicle testing.
  - VSI Engineers are skilled in OpenCV, Neural Networks, Python, C++, control theory, robotics (ROS), and real time computing.
- VSI provides thought leadership to companies around the world and is actively involved in some of the most advanced AV development programs.

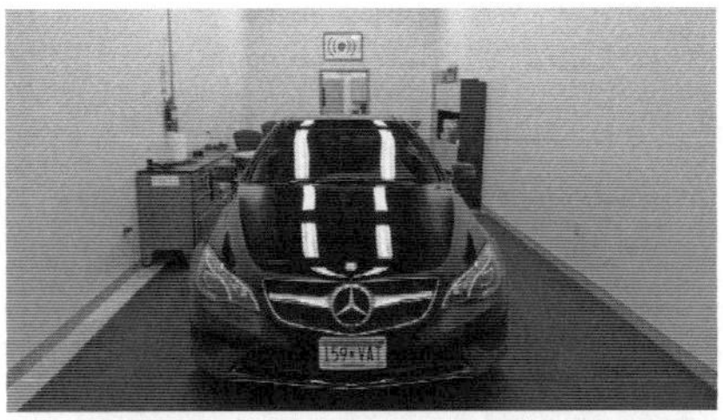

Opening Remarks

- There is a rush to develop the **AV stack** – the HW/SW elements of automated driving. This is being lead by the major chip companies, often in collaboration with tier-ones.
- The AV Stack is a massive collection of IP and is further supported by cloud assets, and a lot of them!
- There is a new round of development platforms for AV technologies – open source, open interface, and SDKs to entice developer activities and build up eco-system partners.
- AV technologies are being targeting by the tech giants (nothing new) – but, they don't plan on building cars... rather they want to control the brains that drive automated vehicles.
- Incremental AV is happing rapidly and in production now... Who will build a better Tesla?
- Meanwhile, Level 4 development is well under way for share mobility platforms... Commercial vehicles operating in urban environments.

VISION
SYSTEMS
INTELLIGENCE

The AV Stack

Developing Automated Vehicle Systems – The "AV Stack"

The AV Stack
Autonomous Vehicle Domain

Application Development Platforms

Run Time SW Components & Basic SW Services

Perception
- ASIC/FPGA
- GPU Accelerators SIMD Memory
- Sensors
- Detectors/CNNs

Localize & Plan
- Data Path Processing
- DDR Memory
- Localization datasets
- Inertia Measurement
- GNSS

Decision / Behavior
- Motion Engine
- ECC Memory
- Behavior Modules
- Arbitration
- Predictors/RNNs

Control
- Lockstep Processor
- DDR Memory
- Safety Monitors
- Fail Safe Fallback
- By-Wire Controllers

Connectivity & I/O
- RF Processors
- Network switches
- Deterministic Bus
- NAND/NOR Memory
- Data Recording

Developing Automated Vehicle Systems –Platforms

Company	AV Domain Controller	Production Platform	Development Platform	Automation L1-L2	Automation L3-L5	Remarks
Audi	zFAS	x			x	SOP* - Audi A8
Nvidia	DrivePX2 (used in Tesla AV 2.0)	x	x		x	SOP* Tesla AP 2
TTTech	TTA Drive		x	x	x	High-end Platform
Visteon	SmartCore Domain Controller		x	x	x	Concept Platform
Renesas	HADP		x		x	TTA Drive w/Renesas
Renesas	R-Car HAD Solution Kit		x	x		ADAS Solution
Bosch	AI Car Computer	x			x	Nvidia-based
NXP	Bluebox		x	x	x	Full HW Stack
TI / RT-RK	Alpha Platform		x	x		ADAS Solution
Delphi	CSLP Platform w/ Mobileye	x		x	x	Prototype
ZF	ZF PRO AI (passenger/commercial)	x			x	Nvidia-based
Hitachi	Autonomous Self Driving ECU	x		x		ADAS Solution
Intel	GO AV Platform (w/Mobileye vision)		x		x	Full HW Stack

* Start of Production

Developing Automated Vehicle Systems – Software

Company	Product	Sensor Fusion	Detection	Interfacing Abstraction	Decision Path Planning	Processing	Control Kit	Testing Simulation
Nvidia	*Driveworks / Drive PX*		✓		✓	✓		
Elektrobit	*EB Assist*	✓		✓				
	Robinos	✓		✓	✓			
Polysync	*OSCC*						✓	
	PolySync	✓		✓				
TTTech	*TTA Drive/ TTIntegration*			✓				✓
Vector	*vADASdeveloper*							✓
Baselabs	*Baselabs Create*	✓						
Mathworks	*Computer Vision Toolbox*		✓					
	Simulink							✓
TASS International	*PreScan*							✓
Autonomos	*Autonomos Core Framework*			✓	✓			
Xilinx	*logiADAK*		✓			✓		
Intempora	*RTMaps*	✓		✓				✓
	I-DEEP							✓
Ricardo	*Agent Drive*							✓
Comma AI	*Neo / OpenPilot*		✓		✓	✓	✓	
Perrone	*DAK*						✓	

Does not include RTE/BSW or MCAL Layers

VISION
SYSTEMS
INTELLIGENCE

Developing Automated Vehicle Systems

Developing Automated Vehicles -- Defining Operating Domains

1. **NHTSA's HAV policy**, published in September 2016, is a regulatory framework and best practices for the development, testing, and deployment of AVs.
2. Automated vehicle systems must be defined with one or more **Operational Design Domains** (ODD).
3. A SAE Level 2, 3 or 4 vehicle could have one or multiple ODDs – e.g. geo-fenced urban, divided highways, automated parking, traffic jam assist, etc.
4. A L5 vehicle has only one ODD as it can (in theory) work anywhere!
5. AVs should be developed, tested and validated against all sceneries that could happen within the ODDs.
6. Each scenario is backed with a **Object and Event Detection and Response** (OEDR) – details how the AV will handle expected and unexpected events.
7. Each scenario must have a **fall back** plan... what to do when system fails.

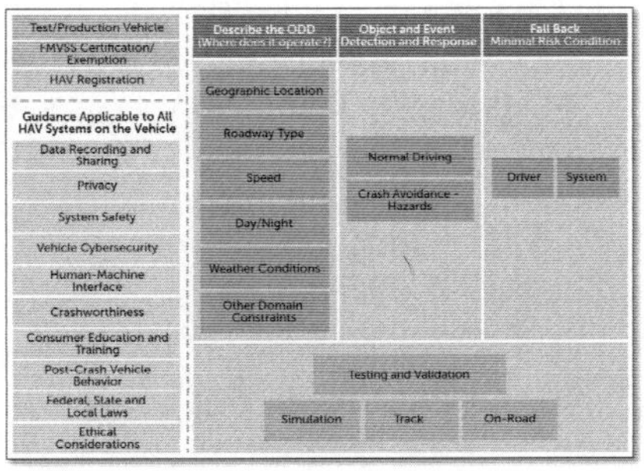

 9

Developing Automated Vehicles – Defining Functional Domains

1. The **perception domain** considers all environment sensors such as camera, radar, LiDAR, ultrasonic, as well as other sensors such as inertia, positioning, throttle position, steering angle etc.
   1. **Fusion of sensor data** can be done with objects (complete sensors modules such as Mobileye) or can be done with RAW data or a combination.
2. The **localization domain** couples knowledge (ground truth) to its environmental model in order to understand its relative position against the real world.
3. The **behavior domain** is where decision are made with respect to motion control.
   1. Active safety systems or L2 automation rely on traditional deterministic algorithms but higher levels of automation will rely on probabilistic algorithms (such as AI) to better predict maneuvers.
4. The **safety domain** (aka Safety Monitor) encases the system as a whole and validates all outputs.

 10

Developing Automated Vehicles – Validation and Testing

- **Simulation** -- allows for the thorough testing of complex systems.
  - These environmental models allow you to define the actors, sensor packages, conditions and scenes
- **Test Tracks** -- Test Tracks allow the creation of some unique scenarios and apply real physics.
  - However, even with excellent test track design and availability, it is impossible to fully test complex systems
- **Field Tests/Real World Data Collection** -- It is important to drive as many miles as possible.
  - Unfortunately, field tests cannot provide enough scenarios (edge cases) to fully qualify a complex system.
- **Artificial Intelligence** – You can use real data or simulated data to train a the Neural Network.
  - To validate, you can examine the layers to see what activates the network... isolate the problems... adjust the weights in the inference model, then test again to check the outcomes.

11

Developing Automated Vehicles – Filling in the Gaps

It is recommended to utilize detailed simulations to develop the systems, test-track tests to validate components and full-vehicles, and field tests to verify the real-life system robustness, whose results can be utilized to train a **neural network** for further testing and simulating of a specific system.

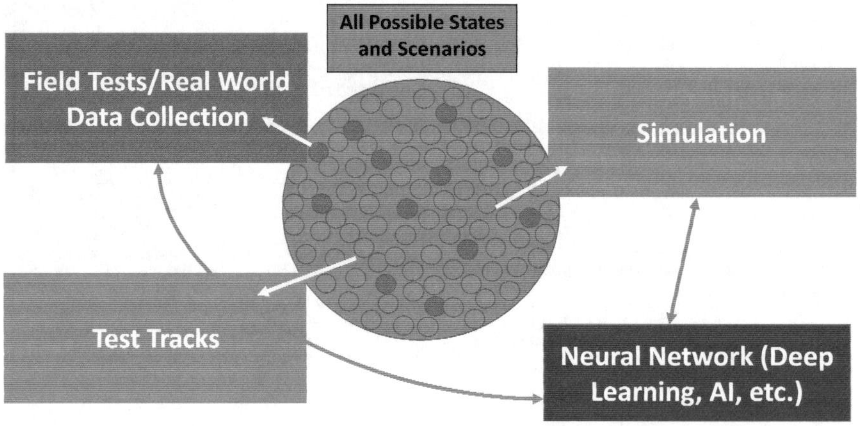

12

Developing Automated Vehicles – Fleet Testing Using *Shadow Mode*

What is it?
- The AV Engine runs in **shadow mode** only – for data collection and evaluation
- Actual AV output commands are recorded alongside real driver inputs
- Once new scenarios are captured, the network is trained and validated, the new feature is then enabled via software OTA

Purpose
- Supplements testing and validation
- Discover new edge cases
- Evaluate machine performance
- Examine human performance vs. machine performance

Tesla is Running Shadow Mode to test and refine features, even before making them available.

Developing Automated Vehicles – Subject Matter Expertise

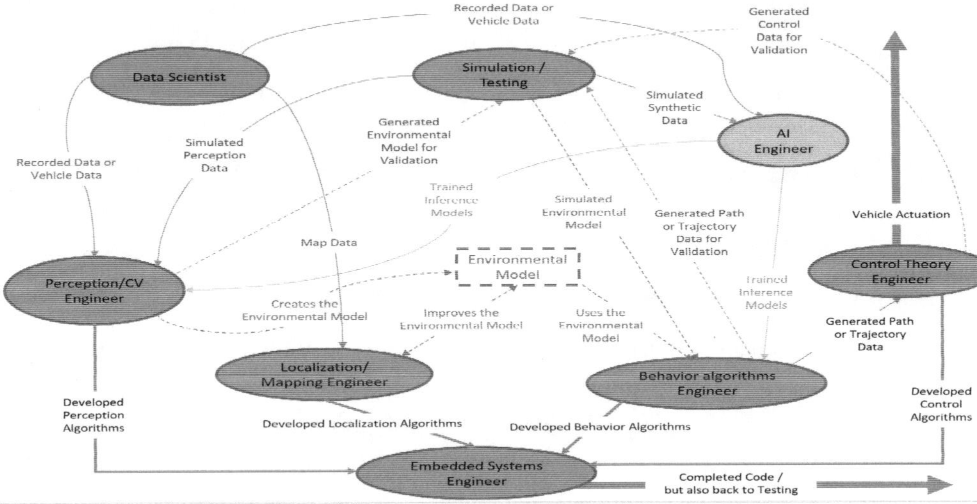

VISION SYSTEMS INTELLIGENCE

The Cloud Assets

 15

Using the Cloud to Manage the Software Stack

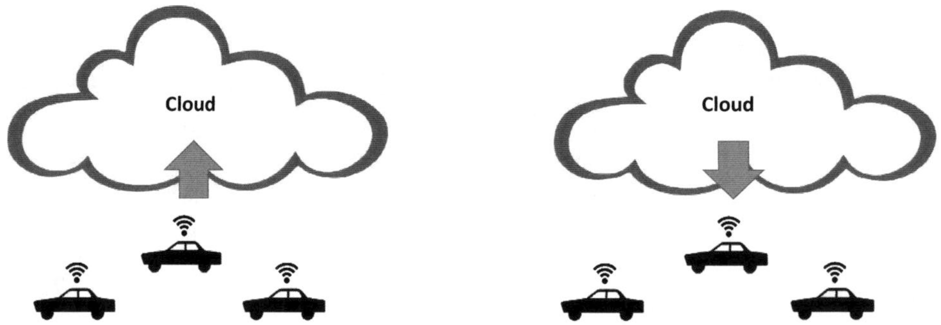

Data Collection

- Record sensor data and uploading the data to the cloud
- Record sensor inputs against human driver output (shadow mode)
- Collect raw data for training the AI-based algorithms
- Collect performance and diagnostic information
- Capture edge cases
- Record objects for localization assets

Data Distribution

- AV Software updates – version control
- Firmware updates to distributed ECU systems
- Updated localization assets – maps and supplemental
- Distributing new inference models
- Distributing fail safe and fall back instructions which will always be changing

 16

Cloud Assets Necessary for Automation

The Internet of Vehicles	
Hardware Device Management	Tight integration of hardware, device drivers, service and board support
OTA Software Enablement	Remote device management through connectivity – layered with security
Developer Eco-System Management	API support for developer community
Software Skills / AI	Artificial Intelligence experience through digital assistants, lots of experience with deep neural networks
Transactions	Handling millions of transactions per day tied to profiles/accounts and financial institutions
Swarms of Geocoded Metadata	Aggregation of user's habits, preferences, movement etc.
Mapping Assets	Map building assets and infrastructure
Data & Analytics management	Ability to handle massive volumes of data for many use cases including the training of AI-based algorithms.

Google
Baidu 百度
amazon
Microsoft
IBM

Closing Remarks

- Incremental automation (L2/L3) is born out of ADAS and becomes a big selling factor over the next 10 plus years before fleet automation starts to dent private ownership.
- Tiers gaining a critical role in AV – handling integration, safety and automotive grade requirements. These company are the bridge between the mechanical and digital worlds.
- OEMs will play a vital role if they play their cards right – building the proper vehicles that meet the requirements of the buyers – nothing new here!
- The balance of power in mobility is changing as the service providers are probably at of top of the list in the era of shared commercial mobility.
- Data acquisition including highly detailed mapping and crowd sourced dynamic content becomes vital to autonomy and to "learning." This will be an important pillar within the AV stack!
- Understanding Operational Design Domains (ODD) is a required specification that explain under what conditions a HAV will operate, how it handles events, and what is the fall back strategy.

VISION
SYSTEMS
INTELLIGENCE

Contact
Phil Magney
Founder and Principal Advisor
phil@visionsystemsintelligence.com
+1-952-215-1797

2016 SCCA ITE Central Divisional Champion

19

AUTONOMOUSTUFF

Enabling the Future of Autonomy

Cameron Gieda
Regional Sales Manager

AutonomouStuff's Goal

OUR GOAL: To enable the future of transportation by significantly reducing development time of autonomy, therefore driving our customers forward

AUTONOMOUSTUFF + YOU =

Vision & Culture

To provide the world's innovative organizations with the ***best*** products, software and engineering services that enable robotics + autonomy. We call this **Bringing Together the World's Best.**

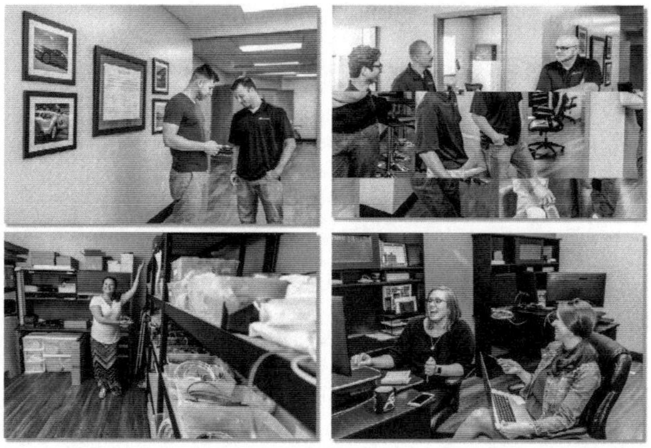

AutonomouStuff is **FULL** of passionate people who love what they do.

Company Profile

AutonomouStuff is a world leader in providing products, engineering services and software that enable *robotics, unmanned off-road systems & self-driving cars*

- Founded in 2010
- More than 2,000 customers worldwide
- Aggressive, continuous growth
- Headquartered in Morton, Illinois
- Strong presence in Silicon Valley, Detroit and China
- World-wide distribution
- Sensor and autonomous drive application experts

Our Customers

AutonomouStuff's customers are in a wide range of industries and *we support them all!*

- Automotive toolchain
- Agriculture
- Mining
- Military
- Aerospace
- Startups
- Maritime
- Railroad
- Universities

Our Customer's Applications

AutonomouStuff's customers provide all sorts of solutions and *we support them all!*

- Autonomous driving cars, trucks, shuttles, boats, security and more
- Convoy/road-train
- Robotics
- 3d mapping
- Object tracking and classification
- Security and safety
- Remote control and monitoring
- Artificial intelligence
- and more

Featured In ...

WSJ · NVIDIA · PCWorld · theguardian · Bloomberg

inside unmanned systems · recode · TechCrunch · autoblog

REUTERS · UST · EETimes · techradar

CAR AND DRIVER · THE VERGE · Make: · JournalStar

... and more

AutonomouStuff Provides ...

Platforms

Our platforms are revolutionizing the future of transportation.

Products

We offer the best technologies so you remain at the forefront of the industry.

Services

Our engineering and knowledge services help you reach your goals.

Software

Software applications are created by our engineers to suit your needs.

Our Fleet of Autonomous R&D Platforms

AUTOMOTIVE		NEV	OFF-ROAD
Lincoln MKZ	Ford Fusion	Polaris GEM	Polaris Ranger

Provided world-wide in 2016-2017
USA, Canada, Germany, Austria, Israel, Korea
CHINA!

Not interested in an MKZ or Fusion? Additional, reliable platforms are coming soon.

World-Class Products

Products offered include:

- 3D
- Vision
- LiDAR
- RADAR
- Actuation
- Middleware
- Data Storage
- GPS/IMU/FOG
- Perception kits
- Wireless controllers
- Rugged computers/servers
- ... and more

Services

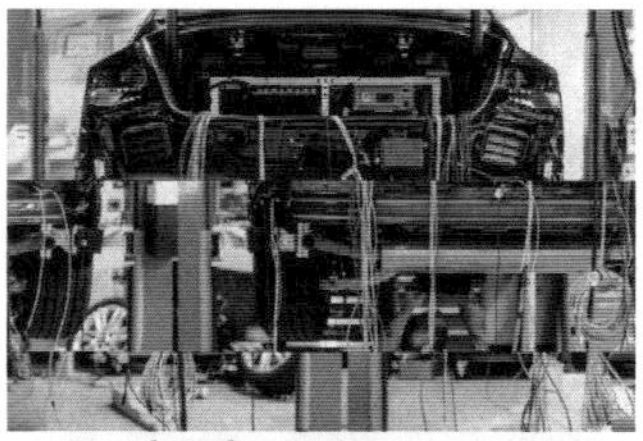

Vehicle Integration

Software Applications

Training

Design & Install

Discovery

Consulting

AutonomouStuff Software Applications

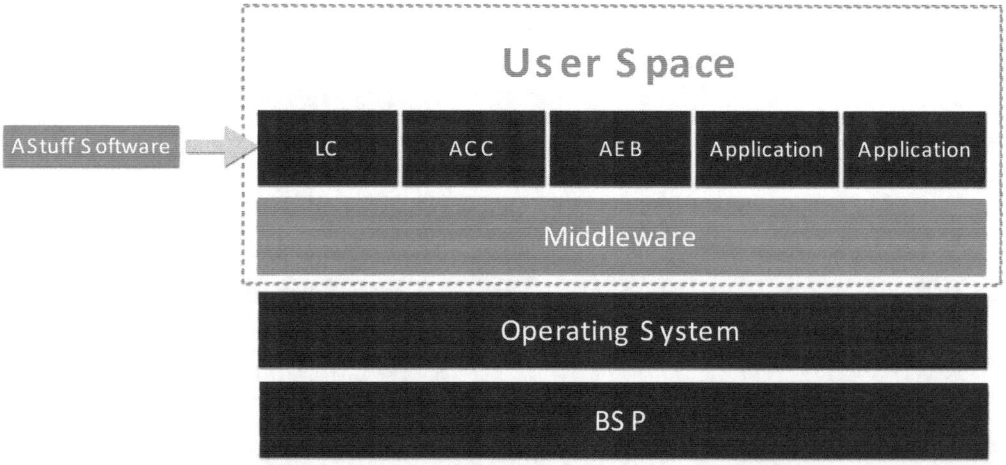

Creating an Autonomous Vehicle

- Autonomous Application
- Vehicle Platform
- Perception/Positioning Sensors
- Data-Fusion Middleware
- Automation Algorithms

Convenience Features

We've mastered the fundamentals of creating an autonomous vehicle and have leveraged our experience to create custom kits for your convenience.

We are proud to have worked closely with NVIDIA to put together kits to help make your transition to autonomy as simple as possible. The work has been done for you so you can focus on research and development from day one!

DRIVE PX on Wheels

These kits take advantage of NVIDIA's latest end-to-end approach, leveraging their DriveWorks software. NVIDIA DRIVE™ PX is proudly supported by AutonomouStuff's talented engineering staff!

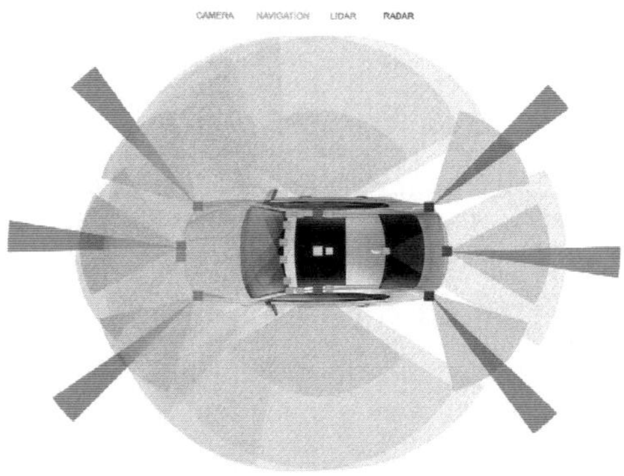

DRIVE PX on Wheels

What does the customer get?

• DRIVE PX mounted in vehicle, pre-wired for cameras, sensors and drive-by-wire kit

• Pre-designed or custom sensor kit

DRIVE PX on Wheels – Advanced Kit

This kit is designed for those who are serious about utilizing the NVIDIA DRIVE PX hardware to its fullest potential.

This is the ultimate NVIDIA DRIVE PX on wheels package!

DRIVE PX on Wheels – Advanced Kit

Kit Includes:

- Vehicle with by-wire kit
- Velodyne VLP16 (2) *VLP 32 coming soon*
- Sekonix cameras (11)
- Xsens MTI-G 710
- RADAR sensors (6)
- NovAtel SPAN IGM-A1
- ibeo 4 Layer LUX sensor
- NVIDIA Drive PX

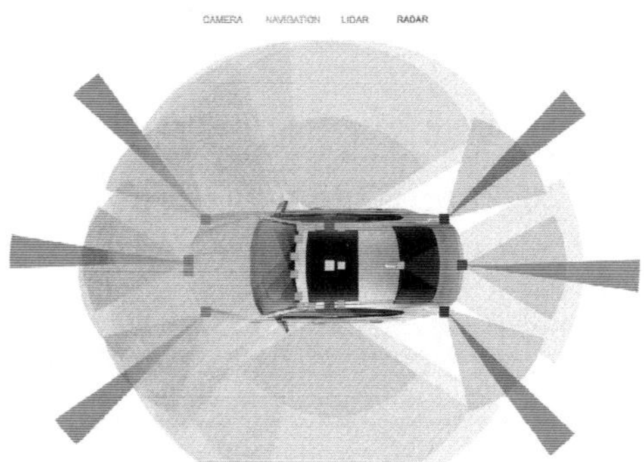

DRIVE PX on Wheels – Basic Kit

This entry-level kit is perfect for those who are interested in becoming familiar with the NVIDIA DRIVE PX hardware on a by-wire equipped Polaris GEM.

A combination of multiple sensor modalities ensures the overlapping field of views from different sensor types.

DRIVE PX on Wheels

Kit Includes:
- Polaris GEM with by-wire kit
- NovAtel SPAN IGM-A1
- Sekonix cameras (5)
- NVIDIA Drive PX

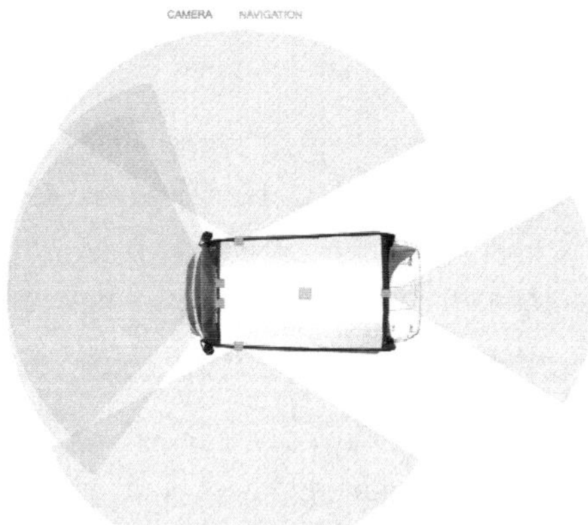

DRIVE PX on Wheels – Custom Kit

We're happy to custom design a kit (and add it to the vehicle of your choice) to meet your specific needs, which can include any combination of our supported kits.

**Subject to design review*

DRIVE PX on Wheels – Custom Kit

Kit Includes:

- Any car
- Velodyne VLP16 *VLP 32 coming soon*
- Sekonix cameras
- Xsens MTI-G 710
- RADAR sensors (6)
- NovAtel SPAN IGM-A1
- ibeo 4 Layer LUX sensor
- NVIDIA Drive PX

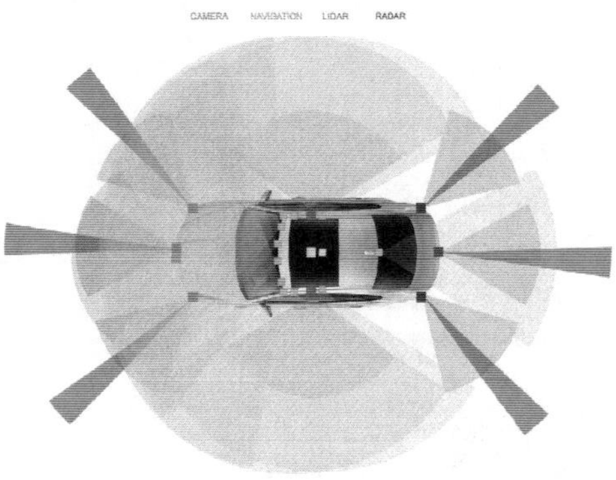

DRIVE PX on Wheels – Support

Join NVIDIA on the NVIDIA Developer Zone for the latest SDK for the NVIDIA DRIVE PX.

If you need assistance on the NVIDIA DRIVE PX, join us on the NVIDIA Developer Forum and let the NVIDIA Community help you out.

NVIDIA & AutonomouStuff

Accelerating Autonomous Driving

Cameron Gieda
Regional Sales Manager
T: 309.291.0966 | M: 408.355.4969
cgieda@AutonomouStuff.com | www.AutonomouStuff.com

Follow us on Twitter! @AutonomouStuff

The National Academies of
SCIENCES • ENGINEERING • MEDICINE

TRANSPORTATION RESEARCH BOARD

AUTOMATED VEHICLES SYMPOSIUM 2017
USERS. VEHICLES. INFRASTRUCTURE.

Neil Pedersen
Executive Director
Transportation Research Board
The National Academies of Sciences, Engineering, and Medicine

Welcome!

About TRB

- Part of the National Academies of Sciences, Engineering, and Medicine
- Private, non-profit institution
- Independent, non-partisan, objective

Research – Convene – Advise

The National Academies of
SCIENCES • ENGINEERING • MEDICINE

TRB
TRANSPORTATION RESEARCH BOARD

TRANSFORMATIONAL TECHNOLOGIES: Transportation

Connected/automated vehicles, shared use services, NextGen, unmanned aerial systems, smart cities, internet-of-things, big data, cybersecurity, alternative-fueled vehicles, 3D printing

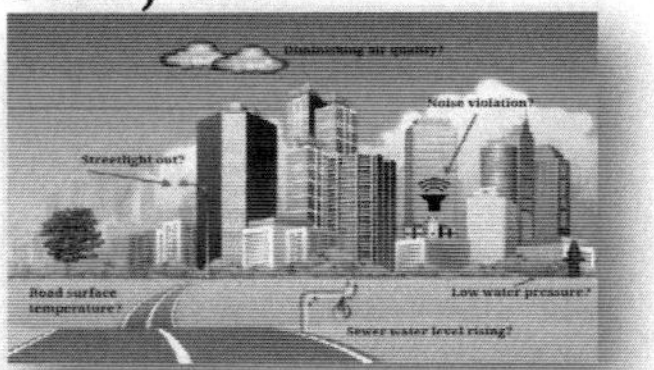

The National Academies of
SCIENCES · ENGINEERING · MEDICINE

TRB
TRANSPORTATION RESEARCH BOARD

Managing Research

- TRB administers over $58M annually in three cooperative research programs
  - Highways & Streets (NCHRP)
  - Transit (TCRP)
  - Airports (ACRP)
- Focused on practical research addressing issues faced by infrastructure owner/operators

The National Academies of
SCIENCES · ENGINEERING · MEDICINE

TRB
TRANSPORTATION RESEARCH BOARD

Connected & Automated Vehicle Research

Since 2014, the National Cooperative Highway Research Program has invested $5M in CV and AV research and results are becoming available:

- Challenges to CV and AV Application in Truck Freight Operations (NCHRP Web-Only Document 231)
- Advancing Automated and Connected Vehicles: Policy and Planning Actions for State and Local Transportation Agencies (NCHRP Report 845, Sept. 2017)
- Impacts of Regulations and Policies on CV and AV Technology Introduction in Transit Operations (Fall 2017)
- Road Markings for Machine Vision (Fall 2017)

The National Academies of
SCIENCES · ENGINEERING · MEDICINE

TRB
TRANSPORTATION RESEARCH BOARD

AV and CV Research (cont.)

Other topics underway include:

- Dedicating Lanes for Priority or Exclusive Use by CVs and AVs
- Implications of Automation for Motor Vehicle Codes
- Providing Support to the Introduction of CV/AV Impacts into Regional Transportation Planning and Modeling Tools
- Business Models to Facilitate Deployment of CV Infrastructure to Support AV Operations
- Cybersecurity of Traffic Management Systems

Find out more at Poster 16 this evening

The National Academies of
SCIENCES · ENGINEERING · MEDICINE

TRB
TRANSPORTATION RESEARCH BOARD

Convene

The National Academies of
SCIENCES • ENGINEERING • MEDICINE

TRB
TRANSPORTATION RESEARCH BOARD

Convening & Connecting Opportunities

- TRB Annual Meeting
  - More than 50 sessions on AVs & CVs in 2017
  - Even more in 2018! (January 7-11)
- TRB Webinars
- TRB Conferences & Meetings
  - Automated Vehicle Symposia (w/AUVSI)
- TRB Standing Committees
  - Vehicle/Highway Automation, ITS, Emerging Technology Law, and more!

The National Academies of
SCIENCES • ENGINEERING • MEDICINE

TRB
TRANSPORTATION RESEARCH BOARD

National Academies/TRB Forum: Preparing for AVs & Shared Mobility

Objective: Academies/TRB Forum on AVs & Shared Mobility

Bring together partners to discuss, identify, and facilitate fact-based research needed to deploy automated vehicles and shared mobility services in a manner and timeframe that informs policy to best meet long-term goals, and to share perspectives on these issues.

- The long-term goals include increasing safety, reducing congestion, enhancing accessibility, increasing sustainability, and encouraging economic development and equity.

Join more than 20 organizations as Forum Sponsors, including:

- U.S. Department of Transportation
- State & Local DOTs
    - CalTrans and California DMV
    - Ohio DOT
    - Washington State Transportation Commission
    - New York City DOT
    - Other State & Local DOTs (NCHRP)
- University of Michigan & American Center for Mobility
- Texas A&M Transportation Institute
- University of California ITS
- AAA Foundation for Traffic Safety
- Cubic Transportation Systems
- Econolite
- WSP USA

For more information, contact Mark Norman (mnorman@nas.edu); or Patrice Davenport (pdavenport@nas.edu)

The National Academies of SCIENCES · ENGINEERING · MEDICINE

TRB TRANSPORTATION RESEARCH BOARD

Thank You

- For more information on TRB, visit: www.TRB.org
- **For more information on TRB Transformational Technologies activities, visit: www.TRB.org/TransTech.aspx**
- **To get involved, go to www.MyTRB.org**

The National Academies of SCIENCES · ENGINEERING · MEDICINE

TRB TRANSPORTATION RESEARCH BOARD

NAHSC

Commemorating the 20th Anniversary of the National Automated Highway Systems Consortium Demo '97 in San Diego

NAHSC

DEMO '97

National AHS Consortium
Technical Feasibility Demonstration

Richard Bishop, Kevin Dopart and Steven Shladover

Origins – 1991 ISTEA Legislation

NAHSC

Section 6054(b) of Intermodal Surface Transportation Efficiency Act of 1991:

"The Secretary shall develop an automated highway and vehicle prototype from which future fully automated intelligent vehicle-highway systems can be developed. Such development shall include research in human factors to ensure the success of the man-machine relationship. The goal of this program is to have the first fully automated roadway or an automated test track in operation by 1997."

National Automated Highway Systems Consortium (NAHSC)

Team formed to compete for FHWA cooperative agreement to deliver the AHS program ($200 M from 1994-2002, with 20% cost share)

- General Motors
- Delphi Automotive (Delco Electronics)
- Hughes Aircraft
- Lockheed-Martin
- Bechtel
- Parsons-Brinckerhoff
- Carnegie-Mellon University
- University of California PATH Program
- California Department of Transportation

Post-Demo Roadmap Toward Deployment

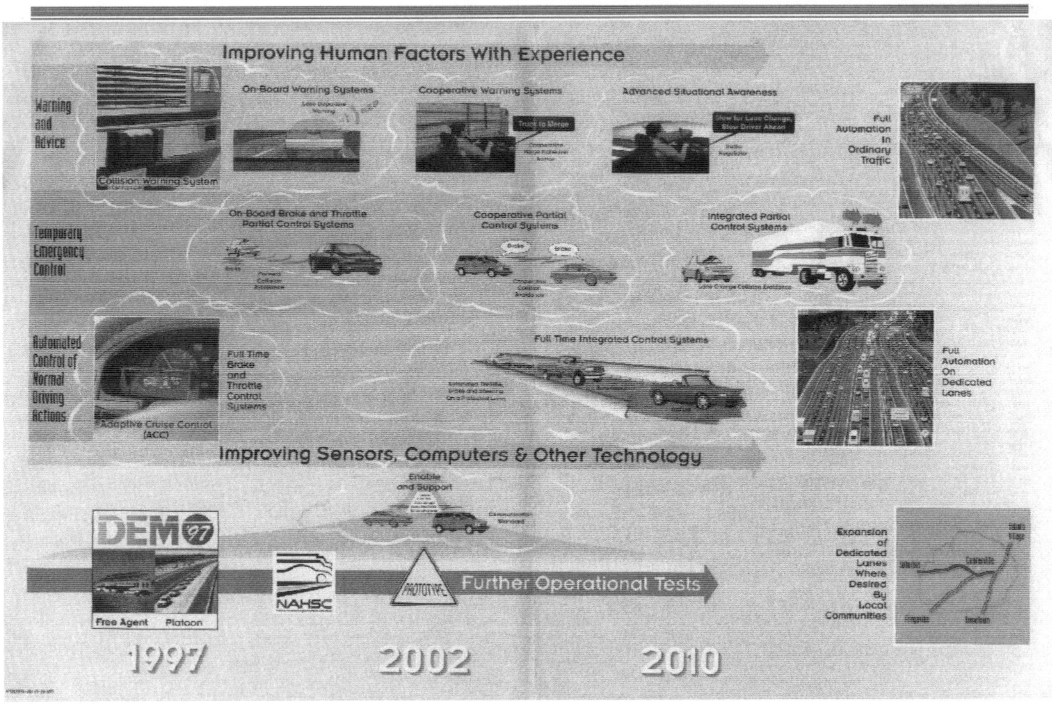

Demo '97 Sites in San Diego

- On I-15 HOV lanes in San Diego and adjacent Miramar College campus (exhibits and mini-demos)
- August 7-10, 1997

Demo Scenarios

- **NAHSC Core Team Demos:**
  - Free Agent Multi-Platform Demo (CMU, GM, Delphi, Hughes, Houston Metro) – 2 buses + 2 cars
  - Platooning Demo (PATH, GM, Delphi, Hughes) – 8 cars
- **Associate Demos:**
  - Toyota – 2 cars
  - Honda – 2 cars
  - Ohio State University – 2 cars
  - Eaton-Vorad – 1 truck (warning) + target car
  - Caltrans/U.C. Davis AHMCT – 2 maintenance vehicles

Free Agent, Multi-Platform Scenario

NAHSC

Platooning Scenario

NAHSC

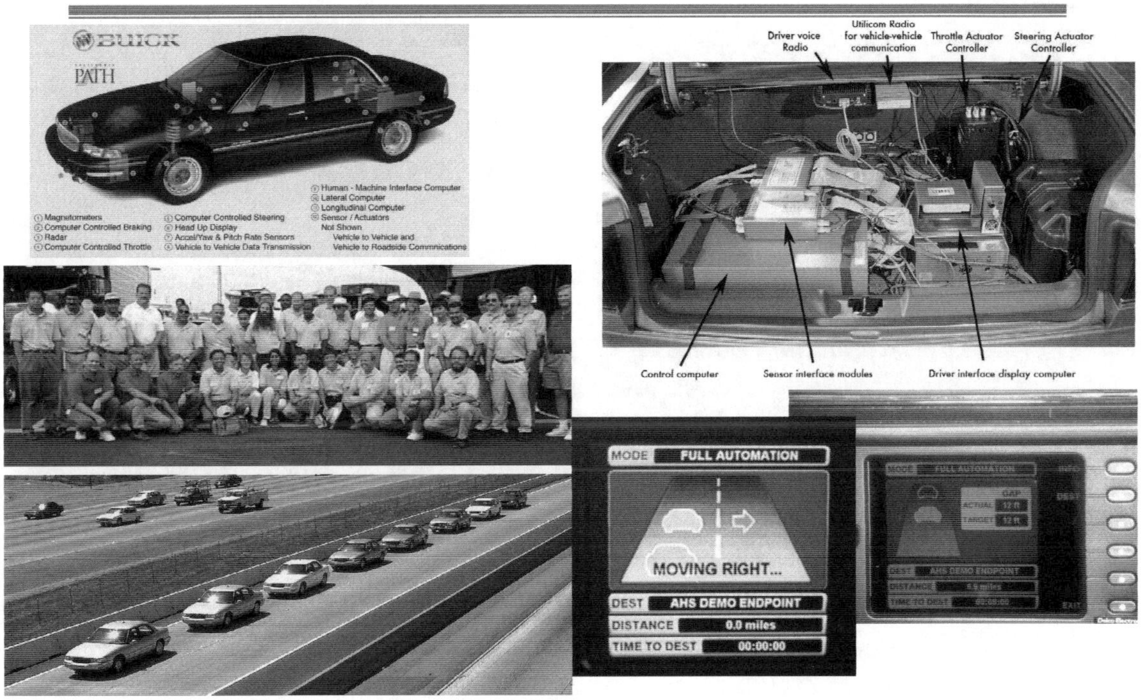

Control Transition Scenario (Honda)

Evolutionary Scenario (Toyota)

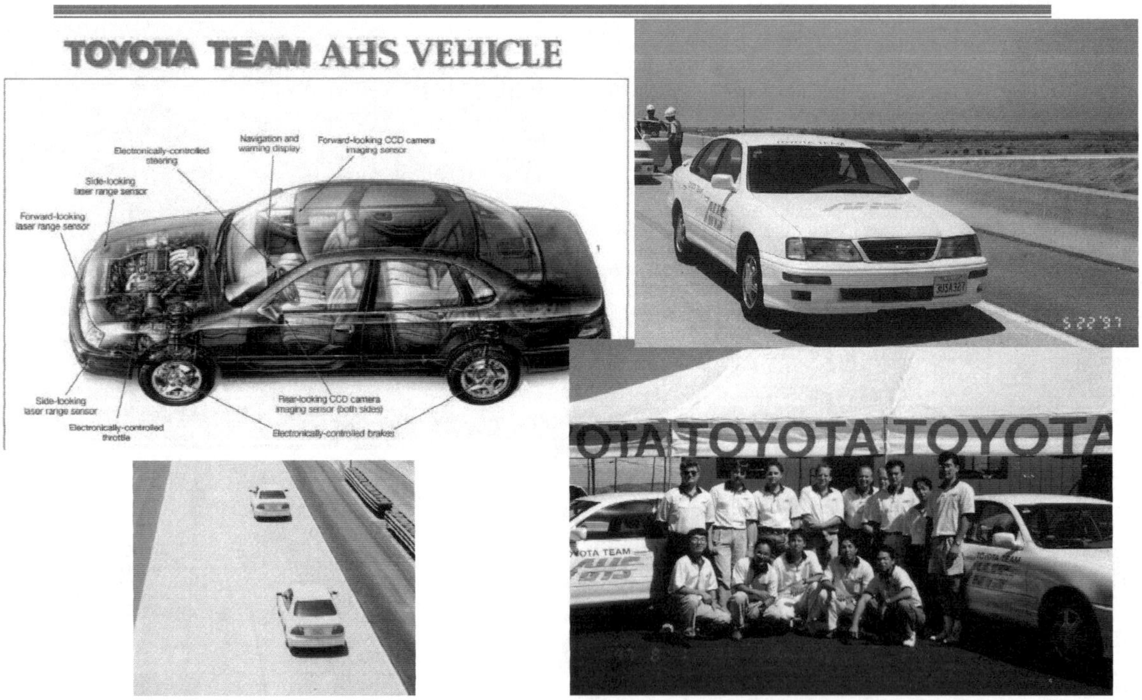

Alternative Technology Scenario (Ohio State University)

NAHSC

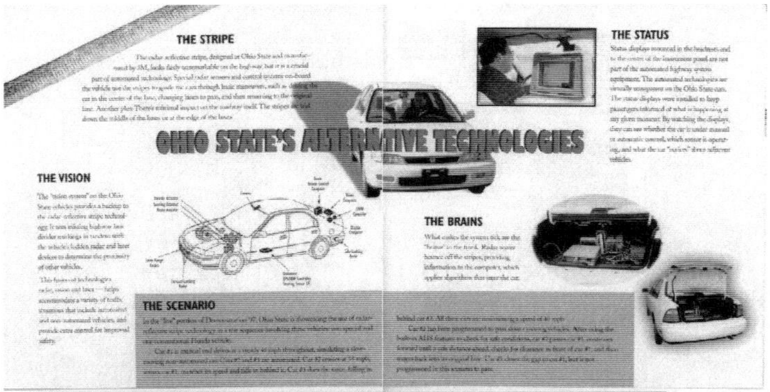

Eaton-VORAD Warning Demo

NAHSC

Caltrans/UC Davis AHMCT Program Highway Maintenance Demo

VIP Visitors and Media Attention

Demo '97 Statistics

- 21 vehicles drove over 8,000 automated miles
- Visitors:
  - Exhibit - 3500
  - Mini-Demo riders: 3100
  - On–road demo riders - 2850
- Media coverage:
  - ABC, CBS, NBC, CNN, MSNBC, PBS television coverage
  - NPR All Things Considered
  - AP, USA Today, NY Times, Washington Post, Wall Street Journal, Newsweek
  - International coverage in 12 countries on 5 continents
  - Total of 120 media outlets represented

Celebration at the End

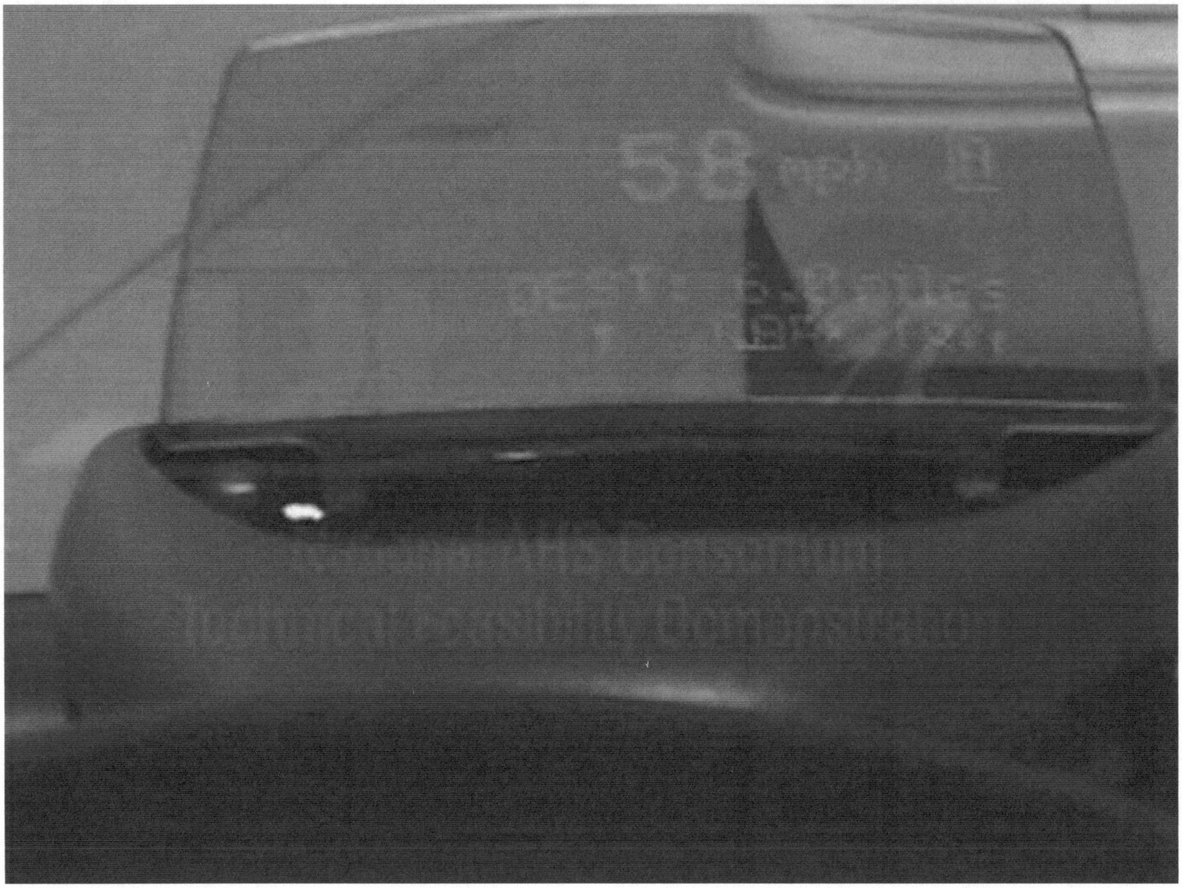

For more information...

NAHSC technical reports and Demo '97 information are posted online at:

http://www.path.berkeley.edu/publications/national-automated-highway-systems-consortium

AVS
2017

LADOT 2016/20167Strategy

Vision Zero + Great Streets

- Improve Project Delivery
- Deliver Great Streets
- Execute Technology Strategy
- Build Community Partnership

Mobility Management

- Develop Parking Reform
- Bring Shared Use Mobility to LA
- Reboot DASH Service
- Evolve and Relocate ATSAC

Great Place to Work

- Pilot Mentoring Program
- Roll out Succession Plan

EQUITY

WHAT IS VISION ZERO?

Reduce traffic deaths to zero by 2025

VISION ZERO
LOS ANGELES | 2015-2025

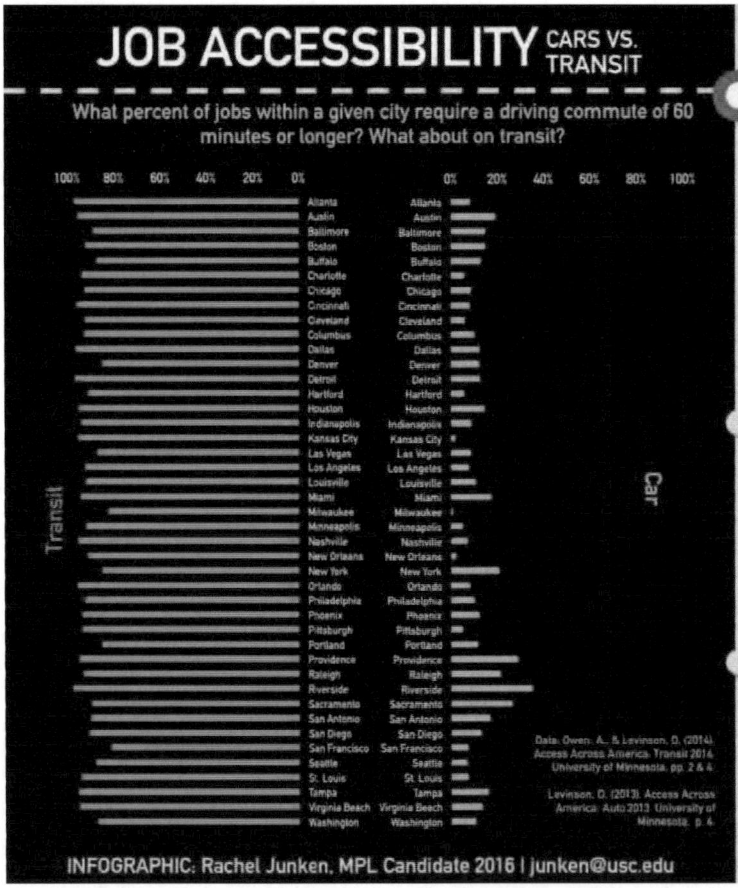

What does this mean for Los Angeles?

92% of jobs require a transit commute of 60 minutes or longer

12x more jobs are accessible by car than by transit in under 60 minutes

7% of jobs require a car commute of 60 minutes or longer

Current and future dedicated-ROW network

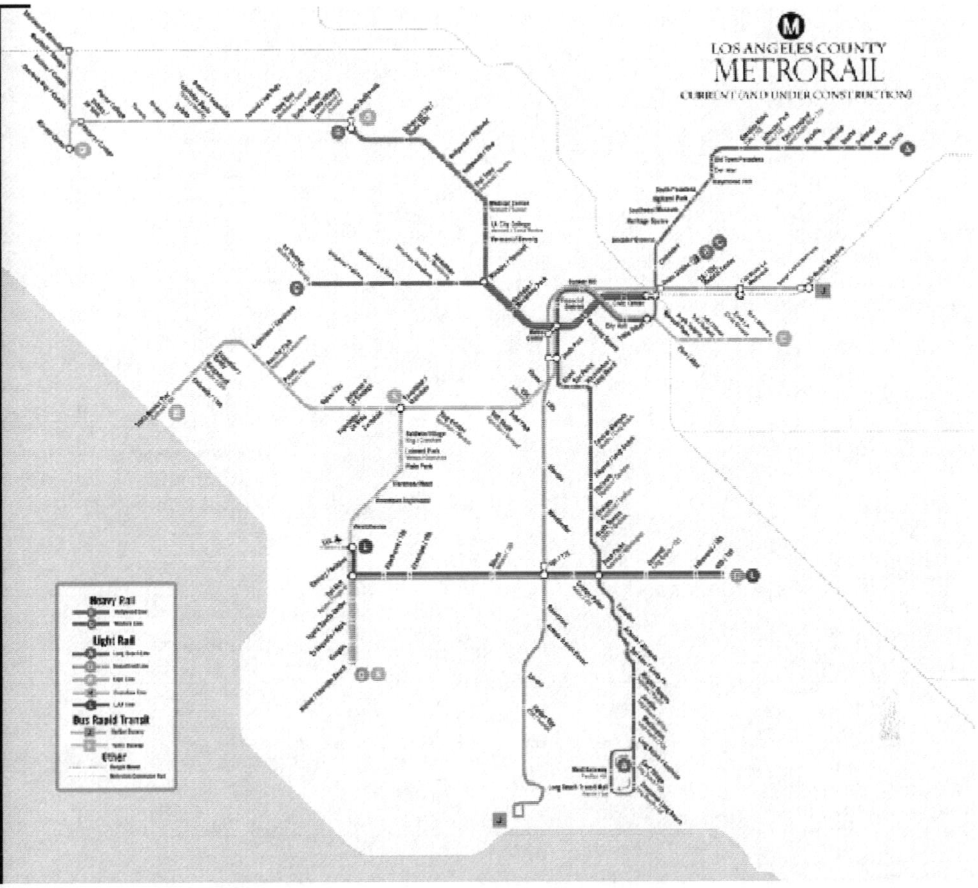

Los Angeles Uber usage map

Spring 2016

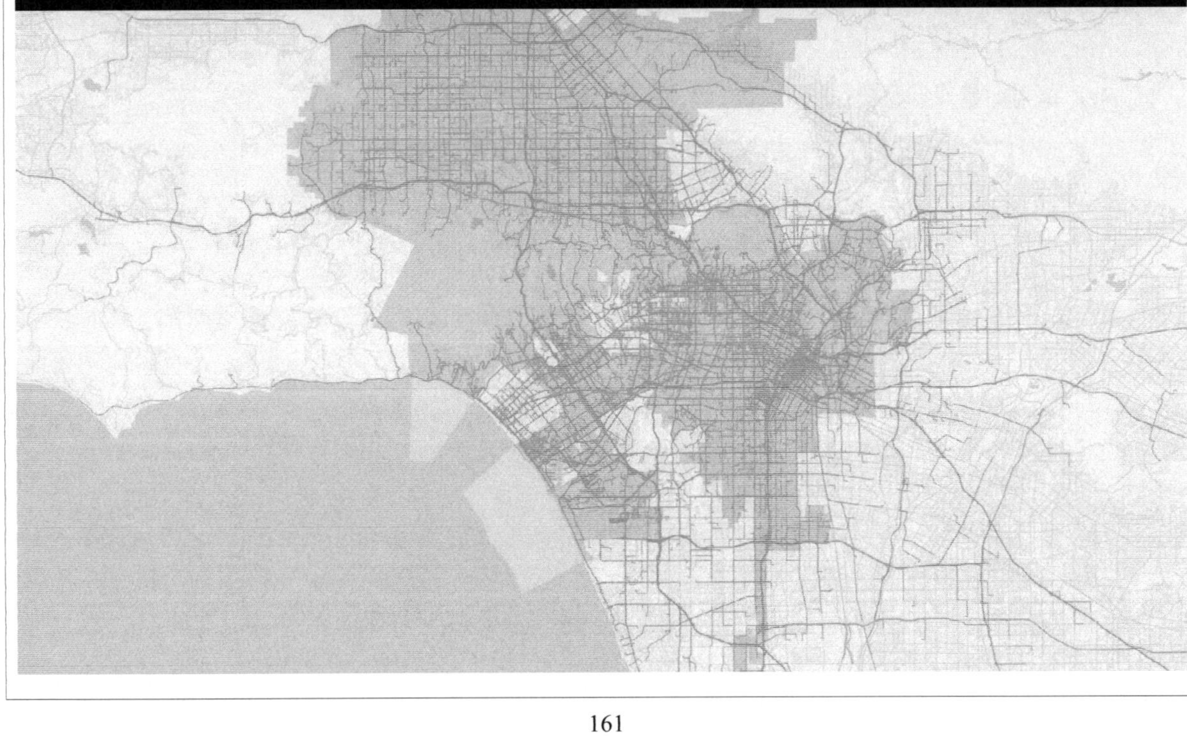

The robots are coming.

+ Autonomous vehicle technology will reshape the transportation infrastructure of Los Angeles

94%	7.5M	1#
Percent of crashes caused by human error	Number of motor vehicles in Los Angeles	Los Angeles's rank among cities with the worst traffic congestion

Source: NHTSA, California DMV, Inrix

8

What if we never widen another roadway?

What if we never build another parking spot?
PASSENGER LOADING ONLY
5 MINUTE LIMIT
SSC 07 CITY OF LOS ANGELES 2710

What if it is just
more of the same?

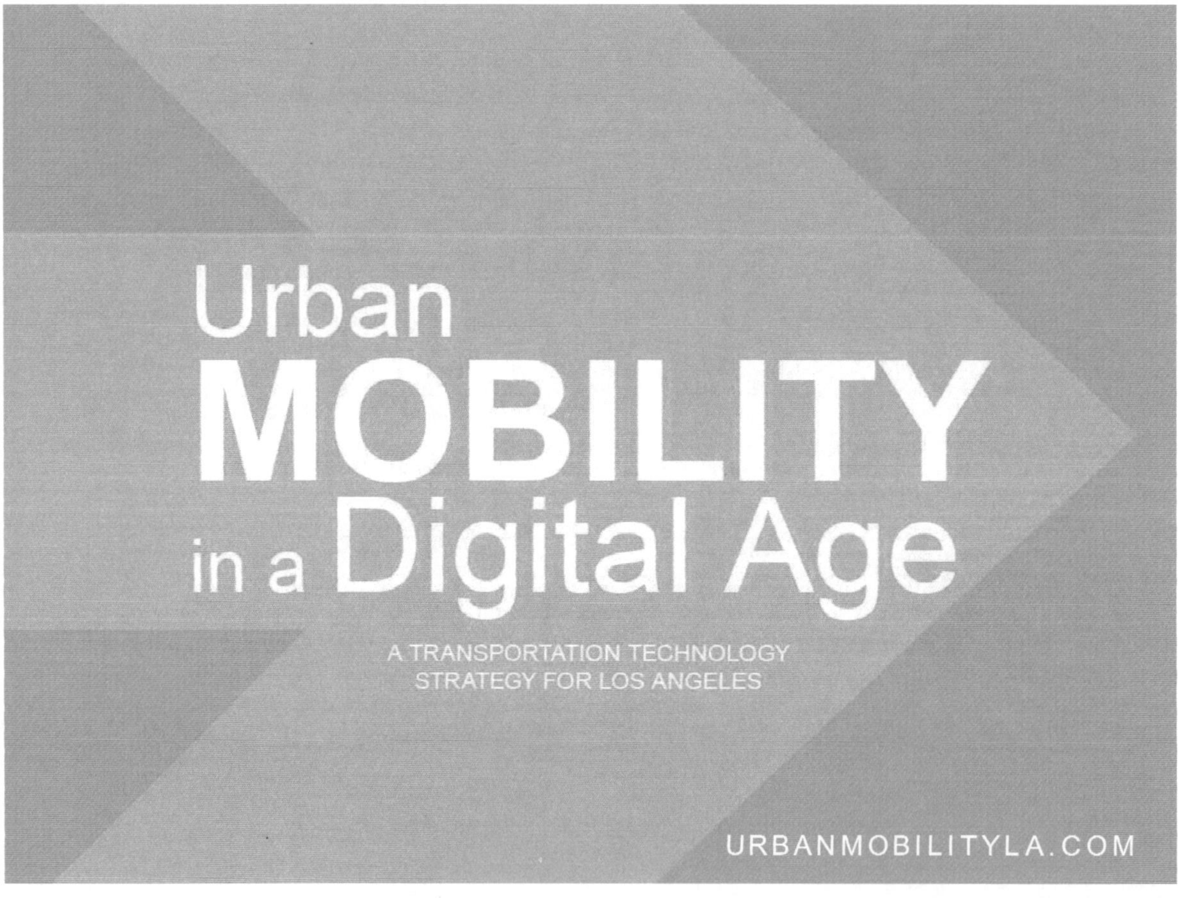
Urban
MOBILITY
in a Digital Age
A TRANSPORTATION TECHNOLOGY
STRATEGY FOR LOS ANGELES
URBANMOBILITYLA.COM

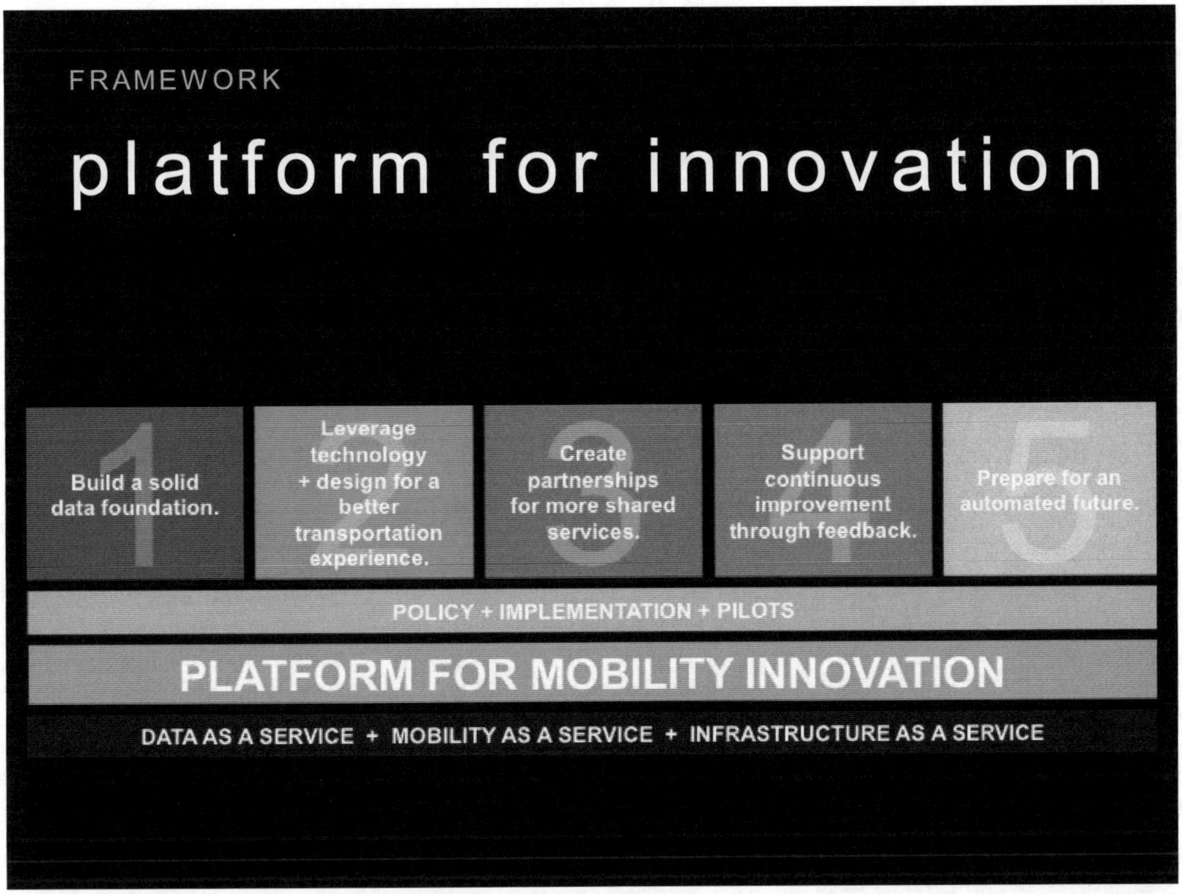
FRAMEWORK
platform for innovation
Build a solid data foundation.
Leverage technology + design for a better transportation experience.
Create partnerships for more shared services.
Support continuous improvement through feedback.
Prepare for an automated future.
POLICY + IMPLEMENTATION + PILOTS
PLATFORM FOR MOBILITY INNOVATION
DATA AS A SERVICE + MOBILITY AS A SERVICE + INFRASTRUCTURE AS A SERVICE

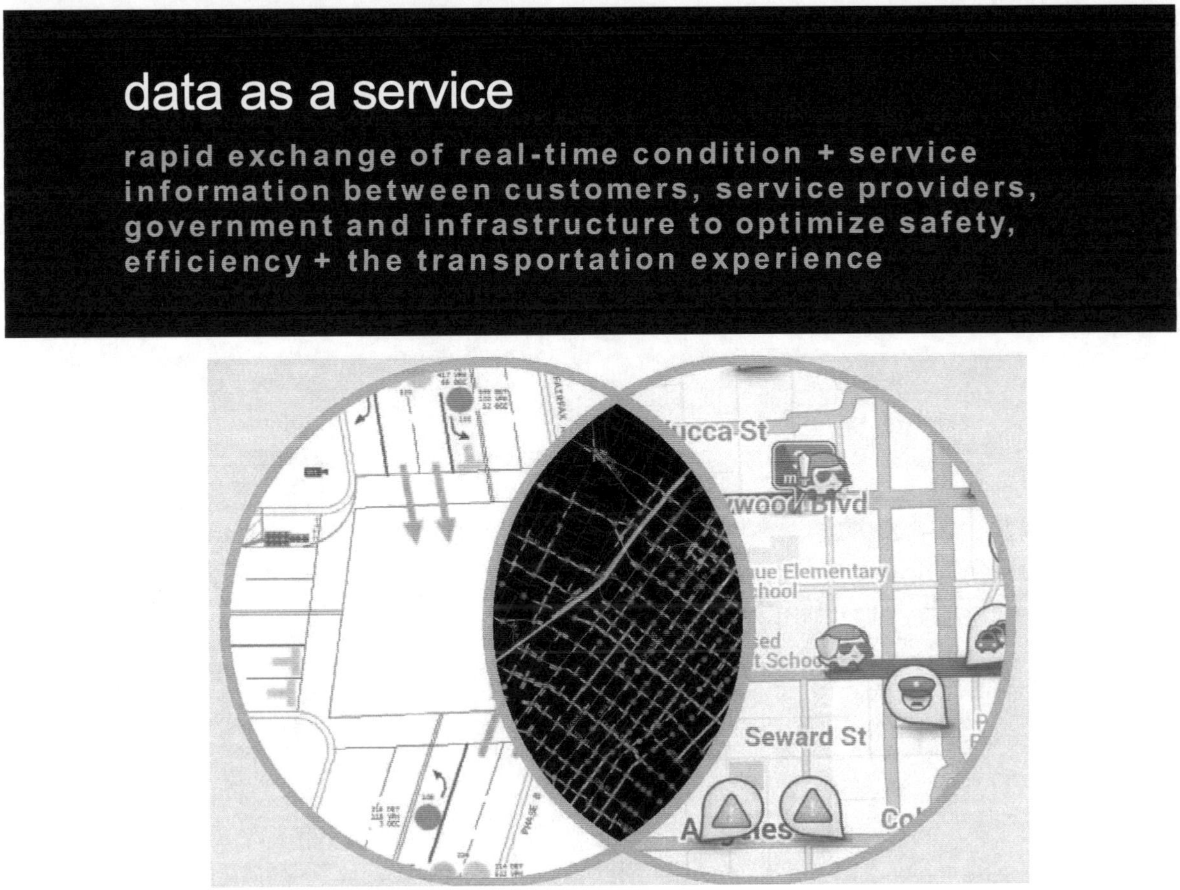
data as a service
rapid exchange of real-time condition + service information between customers, service providers, government and infrastructure to optimize safety, efficiency + the transportation experience
Seward St

Mobility as a Service

access to a suite of transportation mode options through a single platform and payment to simplify access to mobility choices.

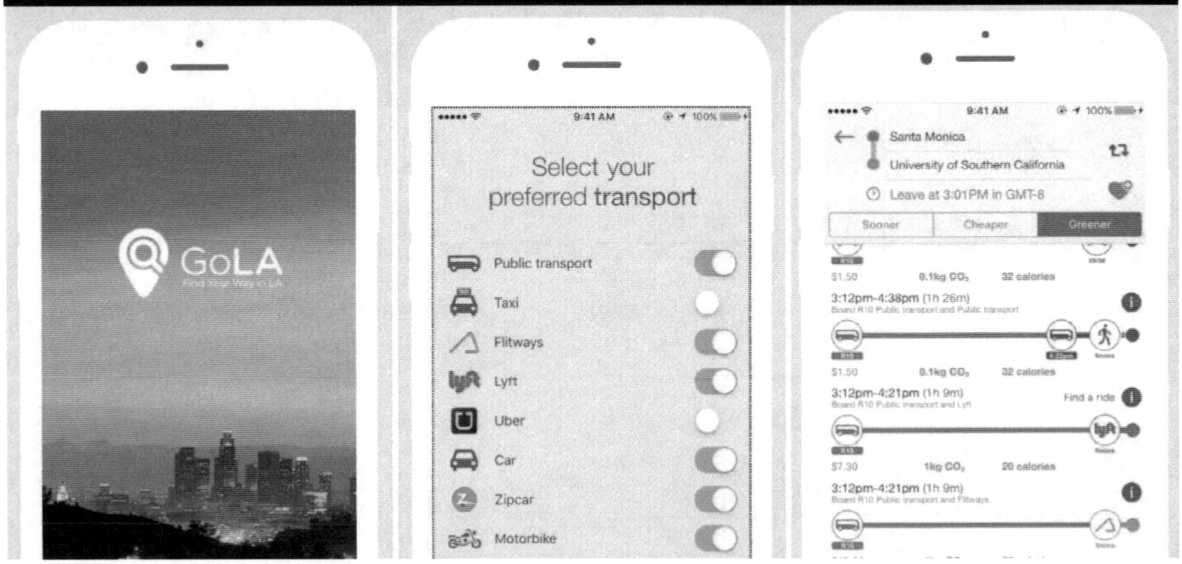

Infrastructure as a Service

A DYNAMIC PAY-AS-YOU-GO APPROACH TO MORE CLOSELY ALIGN COSTS OF PROVIDING INFRASTRUCTURE WITH HOW IT IS IS USED

VISION THE

VISION THEA
(323) 290-238

LA promise zone

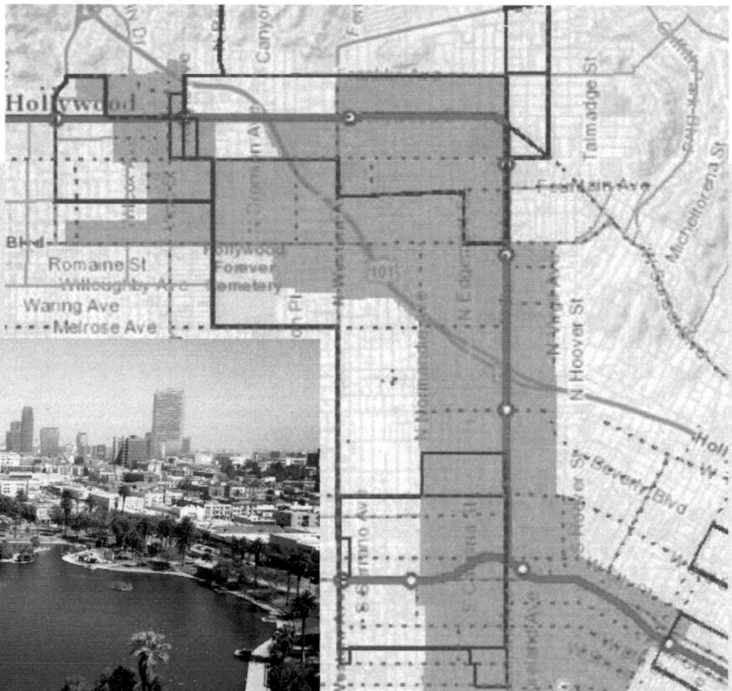

LADOT

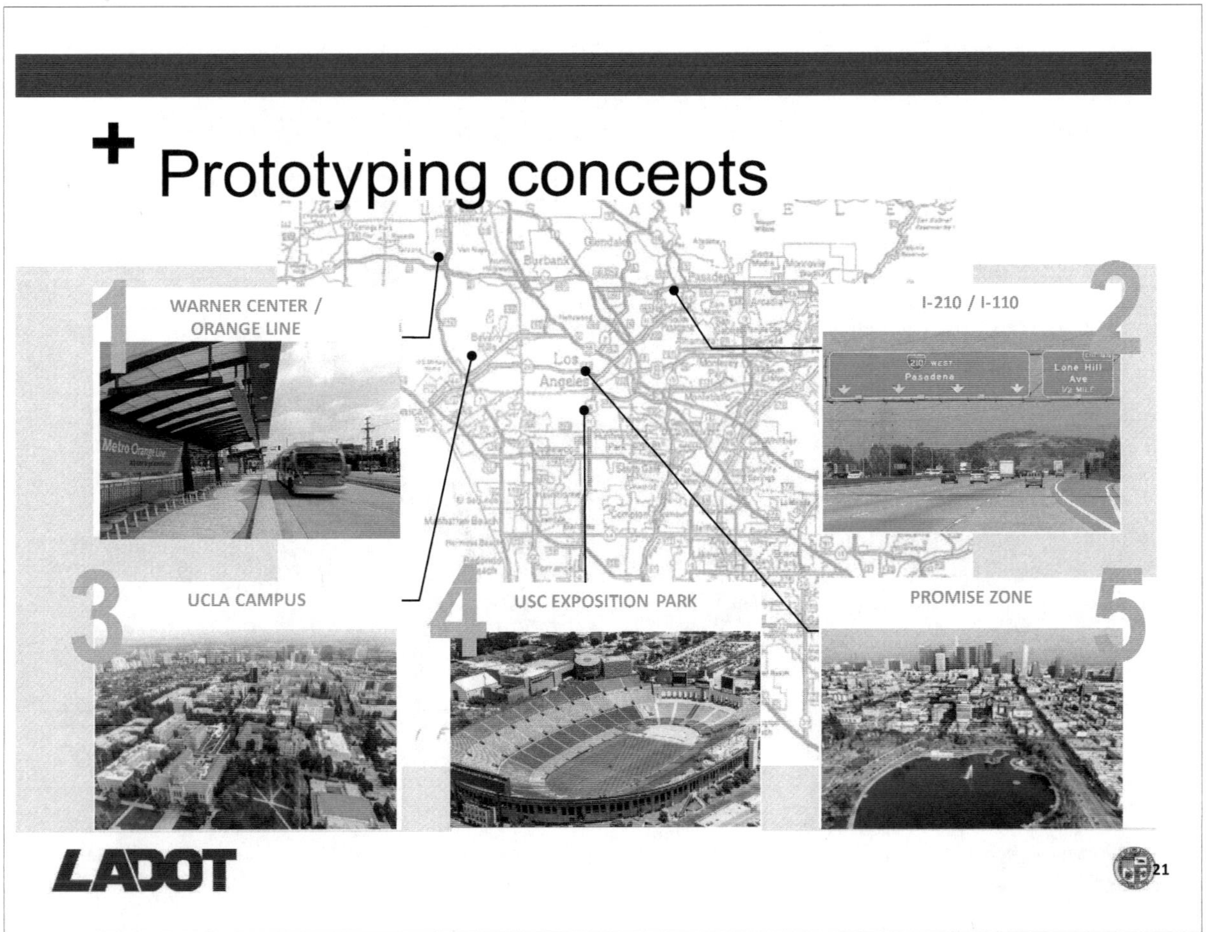

Prototyping concepts
1
WARNER CENTER / ORANGE LINE
2
I-210 / I-110
3
UCLA CAMPUS
4
USC EXPOSITION PARK
PROMISE ZONE
5
LADOT
21

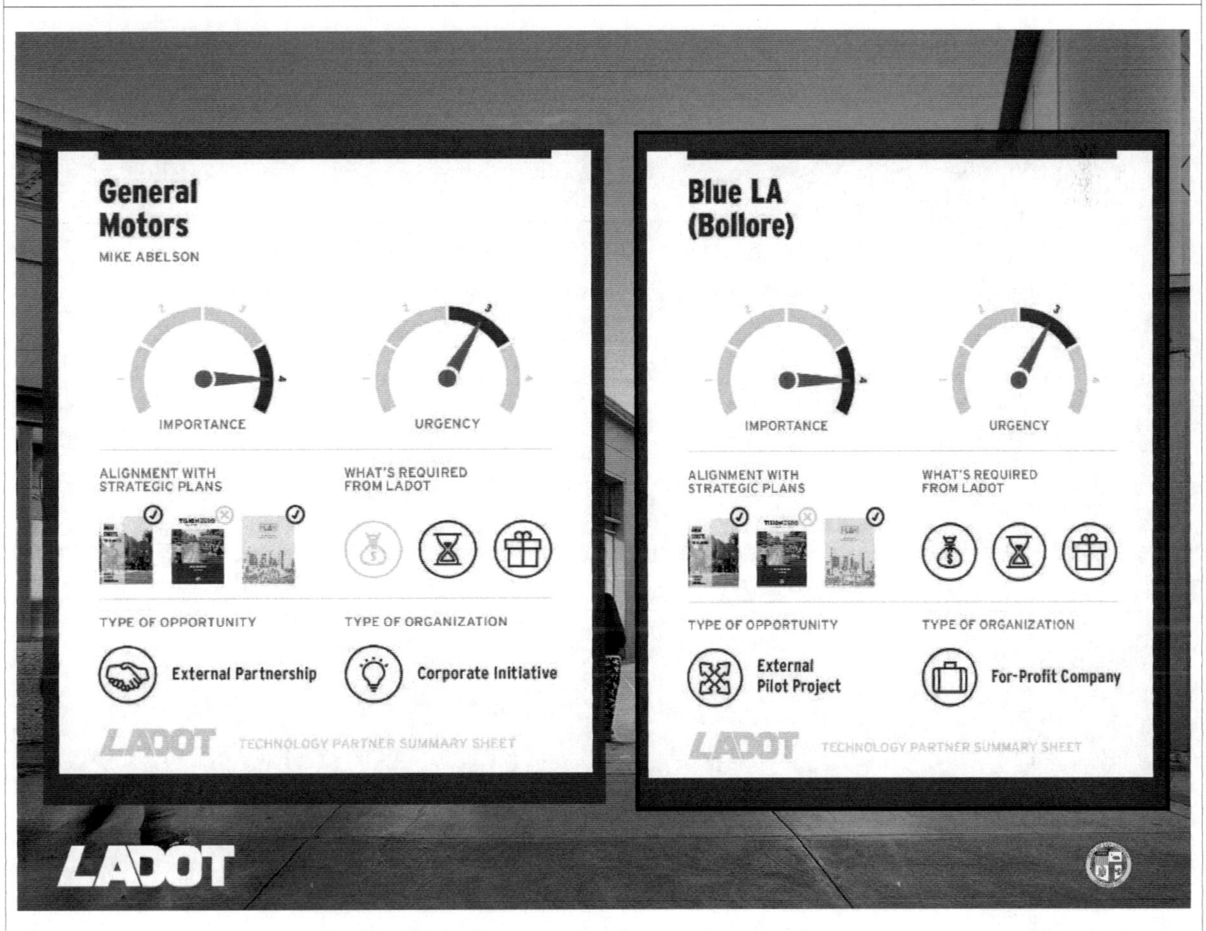

General Motors
MIKE ABELSON
IMPORTANCE
URGENCY
ALIGNMENT WITH STRATEGIC PLANS
WHAT'S REQUIRED FROM LADOT
TYPE OF OPPORTUNITY
External Partnership
TYPE OF ORGANIZATION
Corporate Initiative
LADOT TECHNOLOGY PARTNER SUMMARY SHEET
Blue LA (Bollore)
IMPORTANCE
URGENCY
ALIGNMENT WITH STRATEGIC PLANS
WHAT'S REQUIRED FROM LADOT
TYPE OF OPPORTUNITY
External Pilot Project
TYPE OF ORGANIZATION
For-Profit Company
LADOT TECHNOLOGY PARTNER SUMMARY SHEET
LADOT

+ Tactics

Tactical action steps to engage and co-create

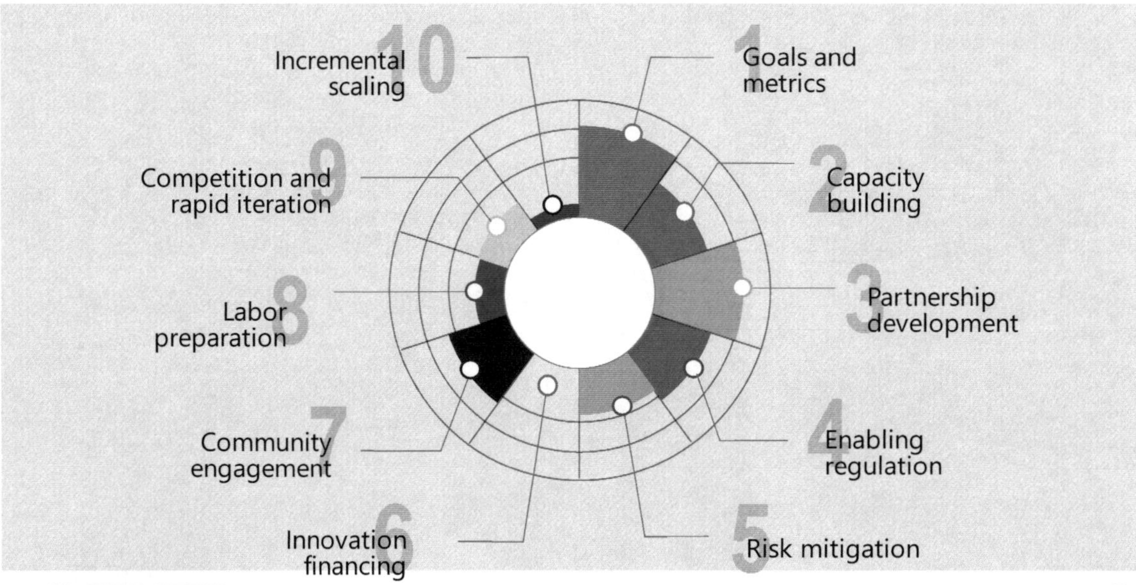

LADOT

Visit visionzero.lacity.org

Email visionzero@lacity.org

@VisionZeroLA

Andrey Berdichevskiy, World Economic Forum

WORLD ECONOMIC FORUM
COMMITTED TO IMPROVING THE STATE OF THE WORLD

Future of Urban and Autonomous Mobility: Bringing Autonomy On and Beyond the Streets of Boston

WORLD ECONOMIC FORUM
COMMITTED TO IMPROVING THE STATE OF THE WORLD

Why does the World Economic Forum care about self-driving vehicles?

Improved road safety

- 90% of accidents today occur due to human error
- Reduction in accidents by 70%[1] feasible if self-driving vehicles represent considerable share of car fleet

Reduced public transport spending

- Reduction in losses from often non-profitable public transport service in lower density areas

Decrease in pollution

- Better fuel efficiency of ~20% can lower overall pollution (absent an increase in mileage)
- Even higher decrease of emissions possible with electrification

Less waiting time

- Seamless, multi-modal end-to-end mobility can be offered to consumers

Freed up space

- Need for parking space in the city can be reduced by up to 60%[1]

Productivity boost

- Over 1.2B hours of pure driving time savings over 10 years possible

Increased traffic efficiency

- Traffic congestion can be improved by ~70%[1] due to smoother traffic flow and fewer cars on the road

Decreased cost of mobility

- Cost savings of up to 50% per km for ride shared self-driving taxi service vs. traditional car ownership

Equitable access to mobility

- Elderly, children and people with disabilities can make use of new end-to-end mobility options

Individual benefits

1. After 10 years; Note: Potential rewards calculated for a model city of ~5M inhabitants; Source: International Organisation for Road Accident Prevention, European Parking Association, UCS, World Economic Forum; BCG analysis

2

WORLD ECONOMIC FORUM

COMMITTED TO IMPROVING THE STATE OF THE WORLD

Multi year Forum initiative to shape the future of urban mobility

2014 — 2015 — Annual Meeting 2016 — 2016 — Annual Meeting 2017

Detailed analysis → **Implementation**

Traffic management
Benchmarking of traffic management concept in leading cities

SDVs: consumer acceptance
Customer research

Self-driving vehicle (SDVs)
Analysis of obstacles in **regulation, society**, and **technology**

SDVs: city perspective
City policy maker interviews

Intermodal travel assistants
Benchmarking of currently available implementations

Future urban mobility scenarios
Scenarios for urban mobility

CITY *of* BOSTON

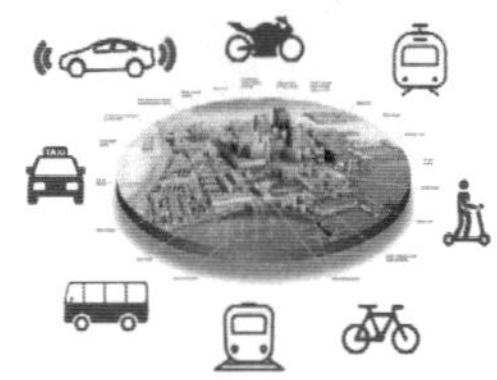

Shape future urban mobility globally in cooperation with the City of Boston

- Develop vision for mobility and strategy for autonomous vehicles (AV)
- Support the set-up of AV testing
- Extend AV strategy to include goods mobility
- Contribute to a network of cities to share best practices and key learnings on the global level

Source: World Economic Forum; BCG analysis

4

The City of Boston realized first tests in 4 months, now three AV partners approved

- **Sep 2016** — CITY of BOSTON, BCG, World Economic Forum: *AV initiative announced*
- **Oct 2016** — CITY of BOSTON: *State and City issue AV executive orders*
- **Nov 2016** — CITY of BOSTON, nuTonomy, massDOT: *MoU signed*
- **Dec 2016** — nuTonomy: *State approves nuTonomy tests*; *Off-road tests completed*
- **Jan 2017** — nuTonomy: *On-road testing kicks off Jan 4th*
- **End Jan 2017** — nuTonomy: *Phase B1 completed*; *100 miles completed*
- **Mar 2017** — nuTonomy: *Phase B2 completed and C1 launched*; *200 miles completed*
- **Jun 2017** — nuTonomy, lyft: *Collaboration announced to work on user interface / experience*
- **Jun 2017** — OPTIMUS RIDE, DELPHI: *Second and third testing partners announced*

Source: World Economic Forum, City of Boston, MassDOT, nuTonomy

5

A phased incubator approach is instrumental to scale technology and business model trials in a controllable environment

Testing phases

- **A** — *Off-site testing*
- **B1** — *100 miles, Marine Industrial Park, day time only, good weather*
- **B2** — *100 miles, Marine Industrial Park, day and night time, mixed weather*
- **C1** — *200 miles, South Boston Waterfront, day time only, good weather*
- **C2** — *200 miles in South Boston Waterfront, day and night, mixed weather*
- **D1** — *400 miles in City of Boston, day time only, good weather*
- **D2** — *City of Boston day and night time, mixed weather*

Detail testing site for phase C1

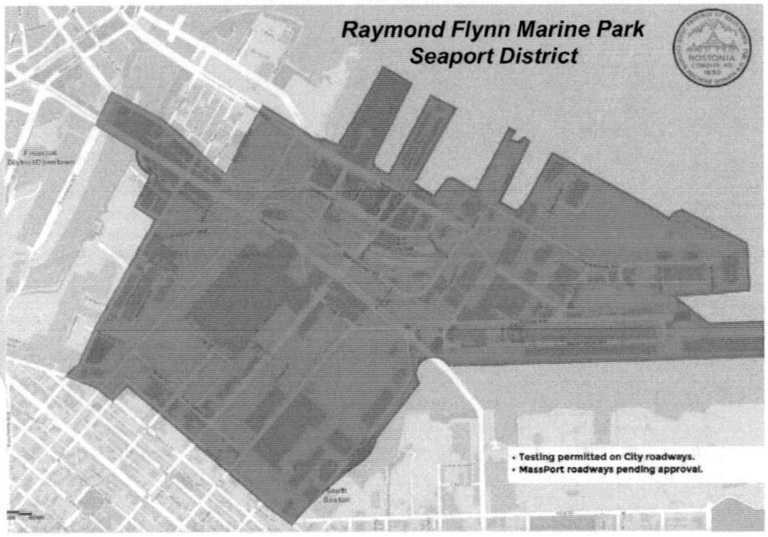

Source: World Economic Forum; nuTonomy, MassDOT, City of Boston, BCG analysis

6

Impact simulation: We built a real-life traffic model for downtown Boston

We took a real world environment incorporating geospatially accurate data ...

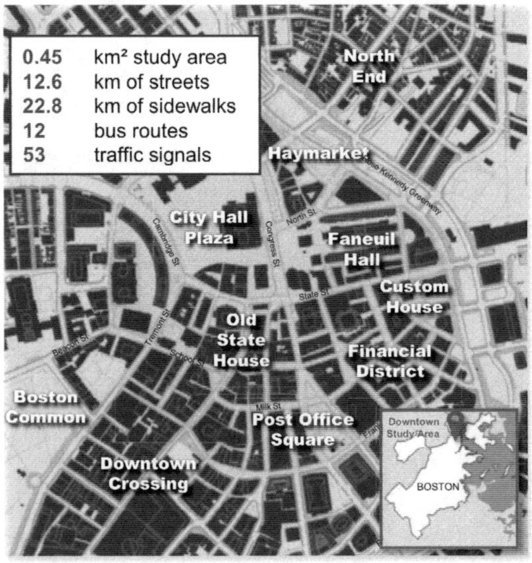

... and simulated traffic flows in its streets

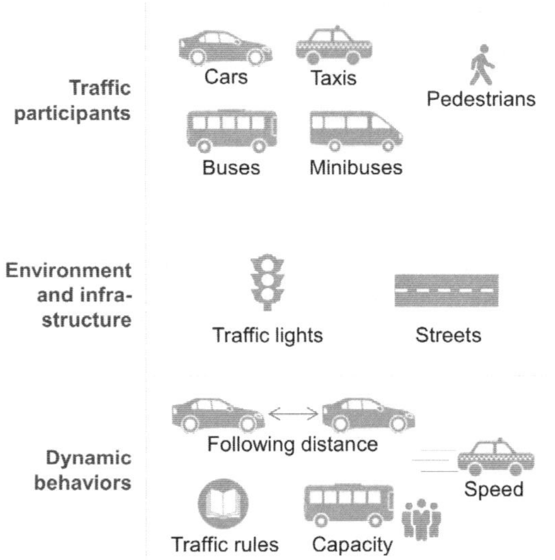

Source: World Economic Forum; BCG analysis in cooperation with MIT Media Lab

7

We looked at one evolutionary and one revolutionary scenario

	Boston today	Private Car Evolution	Robo-Transport Revolution
	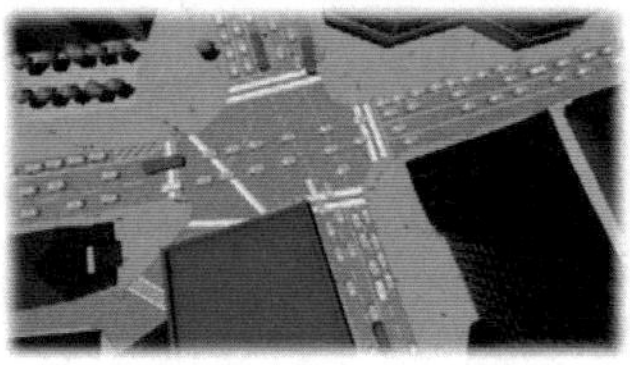	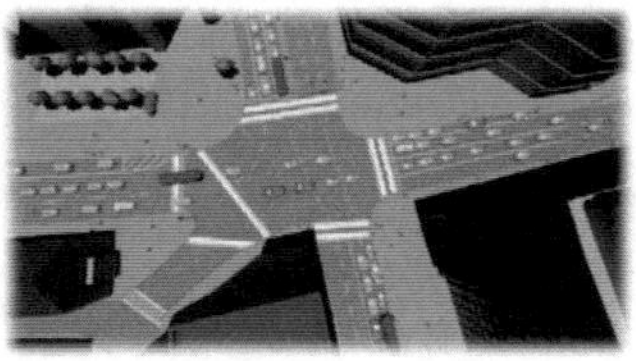	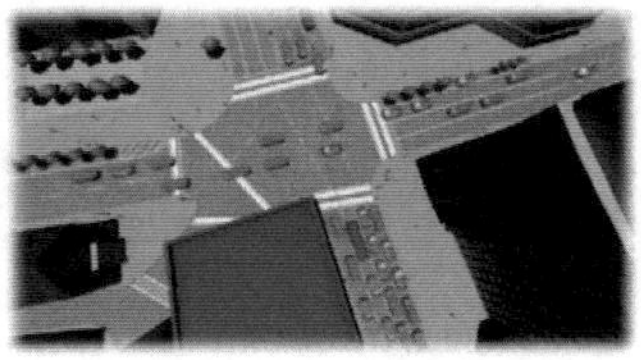
Primary transport modes	*1. Public transit: 56%* *2. Personal car: 33%* *3. Taxi and e-hailing: 11%*	*1. Public transit: 50%* *2. Shared self-driving taxi: 22%* *3. Self-driving personal car: 11%* *4. Traditional personal car: 11%*	*1. Public transit: 34%* *2. Self-driving mini-bus: 28%* *3. Self-driving taxi: 24%* *4. Shared self-driving taxi: 14%*
Scenario description	***Today's status quo in Boston downtown study area*** — Most trips into and out of study area are work commutes — Public transit and personal car as key transport modes	***Shift to autonomous technology with increased sharing*** — Many car owners switch to self-driving cars or using shared self-driving taxis — Some public transit shifts to shared taxi	***Disruptive shift to shared, autonomous transportation*** — Shift from personal car to (shared) self-driving taxi and minibus — Considerable shift from public transit to minibus

Note: Model assumes simplified modal mix without walking and cycling. Boston today modal mix representative of study area only. Modal mix expressed as % of trips taken
Source: World Economic Forum; BCG analysis in cooperation with MIT Media Lab

8

WORLD ECONOMIC FORUM
COMMITTED TO IMPROVING THE STATE OF THE WORLD

Number of vehicles: Considerable reduction due to sharing

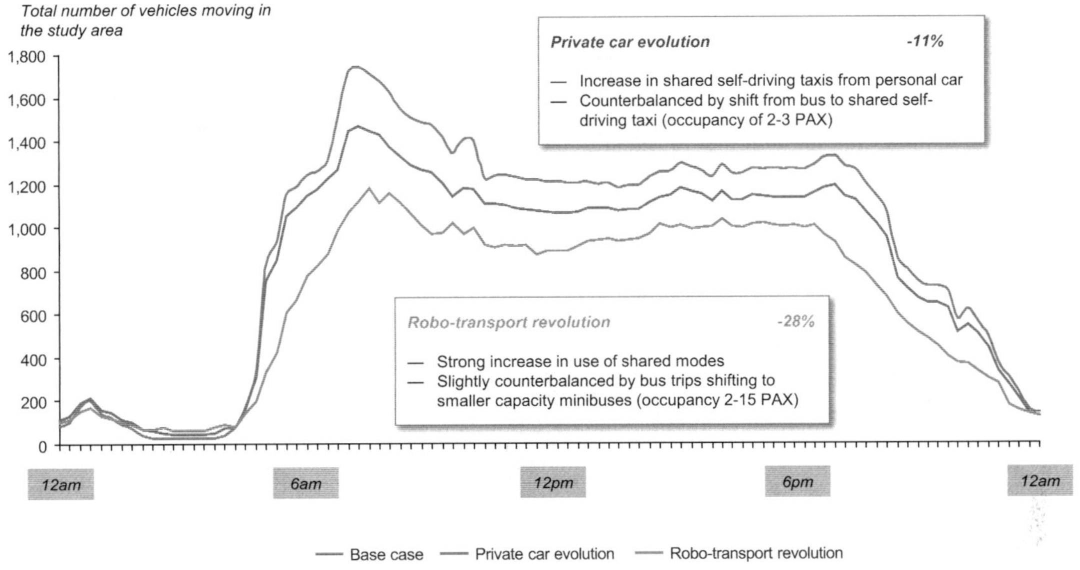

Source: World Economic Forum; BCG analysis in cooperation with MIT Media Lab

9

WORLD ECONOMIC FORUM
COMMITTED TO IMPROVING THE STATE OF THE WORLD

Average travel speed: Significant improvements of up to 30% possible

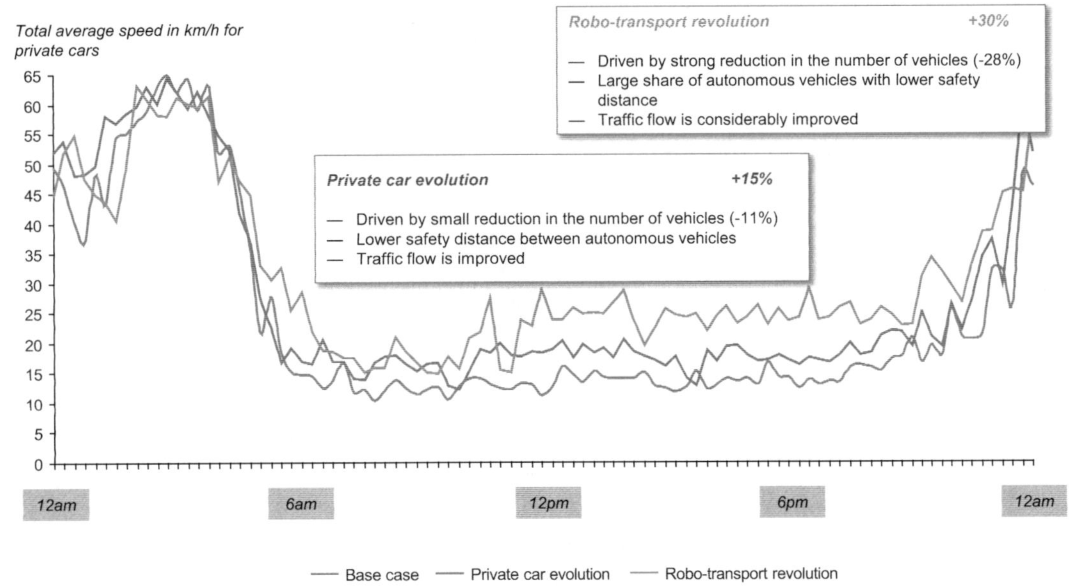

Source: World Economic Forum; BCG analysis in cooperation with MIT Media Lab

10

In both scenarios we found significant impact along key KPIs

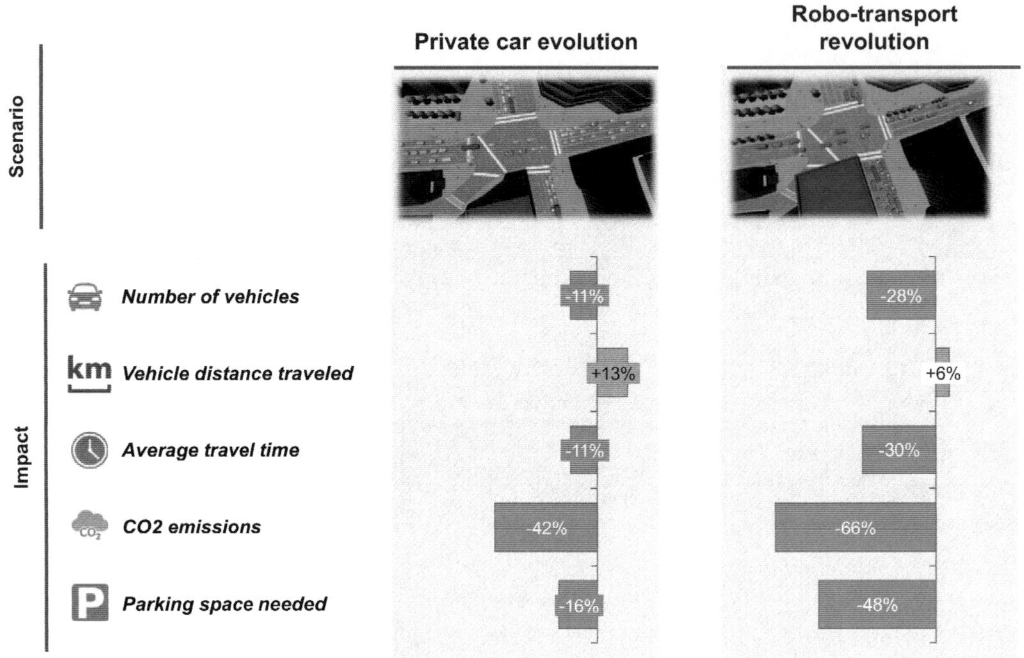

Source: World Economic Forum; BCG analysis in cooperation with MIT Media Lab

11

Six key takeaways from the collaboration with the City of Boston

1. Autonomous vehicles are a crucial building block to make transportation more **accessible, safe and reliable**
2. Autonomous vehicles **enhance, but do not replace public transit**
3. Getting used to AVs takes time—**public's awareness** for them must be **created early on**
4. City of Boston does **not want to own assets** for shared mobility models
5. Boston envisions **one city-wide mobility platform** where all its mobility offers converge
6. **Experimentation with different industry partners** key to learn, always in close cooperation with state level

Source: World Economic Forum; BCG analysis; City of Boston

12

Outlook: The project focusses on 5 key topics in 2017

1 *Urban logistics models*

Analyze new business models for urban goods delivery

2 *AV testing*

Assist in expanding autonomous vehicles testing in Boston

3 *Impact study AVs*

Broaden scope of traffic simulation

4 *Mobility Platform*

Develop framework for city mobility platform

5 *City Network*

Facilitate a best and worst practice on autonomy leveraging digital capabilities and existing city networks & collaborations

13

Contact Information

Andrey Berdichevskiy
Practice Lead Urban Mobility
Global Leadership Fellow
World Economic Forum
Tel.: +1 646 203 2755
E-mail: abed@weforum.org
Twitter: @andreyberd

BACKUP

Autonomous mobility landscape 2016: Variety of categories emerged

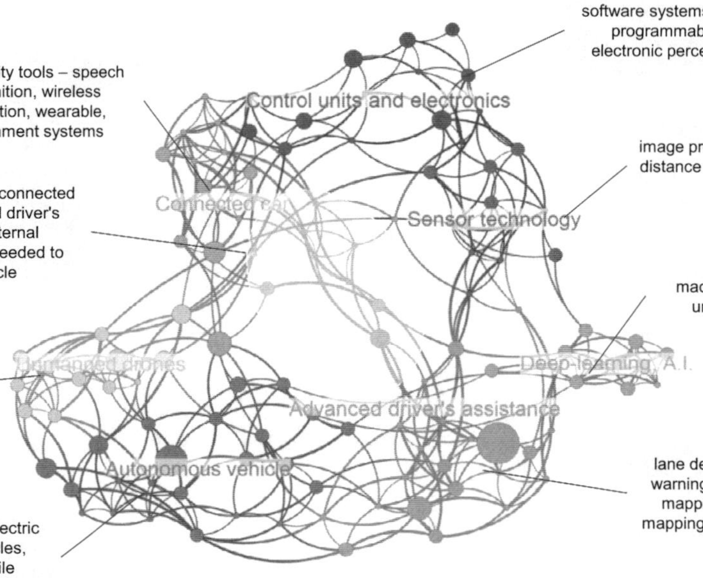

Source: BCG analysis with Quid

15

BACKUP

From control units to autonomous vehicles and AI in ~15 years

2000
33 companies

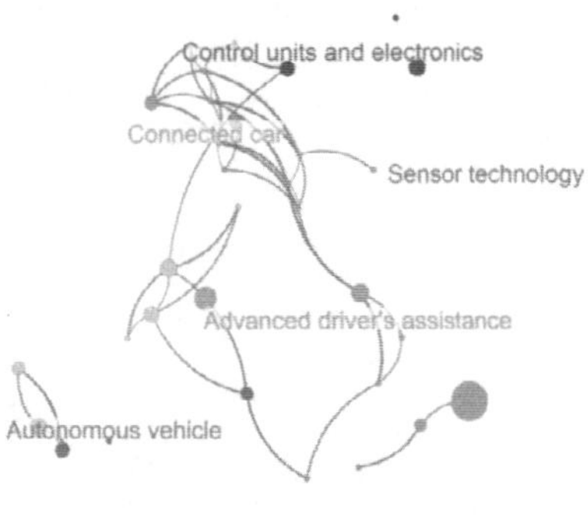

2008
55 companies

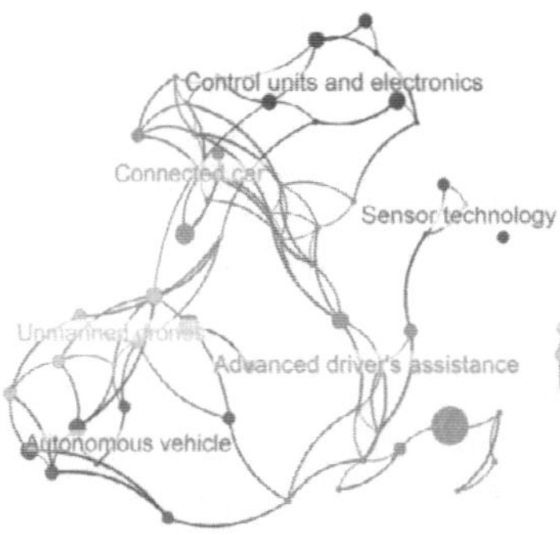

2016
115 companies

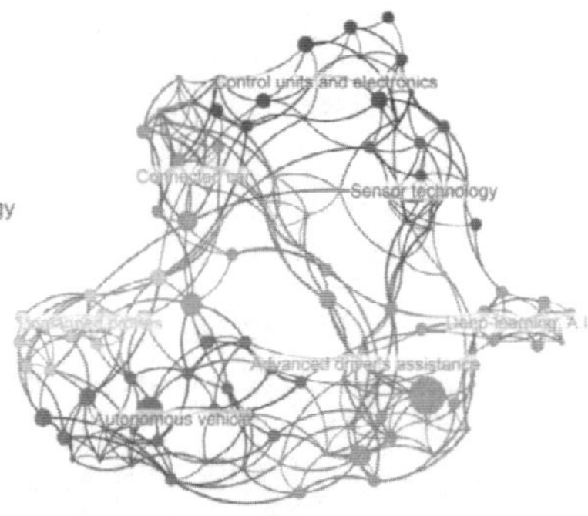

Connected car systems and **Control units** were already in vehicles

Unmanned drones were the earliest form of 'autonomous vehicles'

Deep-learning and A.I. is a completely new cluster. Large growth in **Sensor technology** and **Advanced driver assistance**

Source: BCG analysis with Quid

16

Truck Automation Overview

Automated Vehicles Symposium, Truck AV Panel

July 12, 2017

Richard Bishop

BISHOP CONSULTING

What Factors Affect Trucking Today?

- Driver shortage
- Hours of Service
- Fuel cost
- Crashes
- Congestion
- Sustainability
- Trailer Length / Longer Combination Vehicles
- Increasing home-delivery parcel volumes

AV Use Cases for Heavy Trucks: Driven and Driverless

Open Road

- Fuel Economy
  - Truck Platooning
- Productivity
  - Traffic Jam Assist
  - Highway Pilot
  - Automated Trailer Backing
  - Parcel Delivery Automation

Constrained Environments

- Trailer Switching
- Mine Hauling
- Drayage

BISHOP CONSULTING

AV Use Cases for Heavy Trucks: Driven and Driverless

Open Road

- Fuel Economy
  - Truck Platooning
- Productivity
  - Traffic Jam Assist
  - Highway Pilot
  - Automated Trailer Backing
  - Parcel Delivery Automation

Constrained Environments

- Trailer Switching
- Mine Hauling
- Drayage

BISHOP CONSULTING

BISHOP CONSULTING 5

Near Term: Driver Assistive Truck Platooning

- Two trucks following safely at close following distances
- Significant fuel savings due to aerodynamics
  - 5% for front truck and 10% for rear truck in two truck platoon are typical
- Combining vehicle-vehicle communications with radar
  - braking by front truck occurs simultaneously on follower truck
- Level One system: follower truck driver still responsible for steering
- Gaps variable in real-time based on conditions

BISHOP CONSULTING

Evolution of Truck Platooning

- Level One product introduction imminent
- Within Fleet >> Inter-fleet
- Longer platoons?
- Evolution of driver role in follower trucks to higher levels of automation

BISHOP CONSULTING 7

AV Use Cases for Heavy Trucks: Driven and Driverless

Open Road

- Fuel Economy
  - Truck Platooning
- Productivity
  - Traffic Jam Assist
  - Highway Pilot
  - Automated Trailer Backing
  - Parcel Delivery Automation

Constrained Environments

- Trailer Switching
- Mine Hauling
- Drayage

BISHOP CONSULTING

Port of Palm Beach Gate Congestion

9

Start-Up Approaches

- Business case:
  - safety
  - fuel economy
  - enhancing driver's job
  - reduced cost by replacing driver with automation
- Current concepts
  - highway driverless: Level 4 exit-to-exit
  - street driverless: remote control
  - private property driverless: distribution centers

BISHOP CONSULTING 10

Panelist Introductions

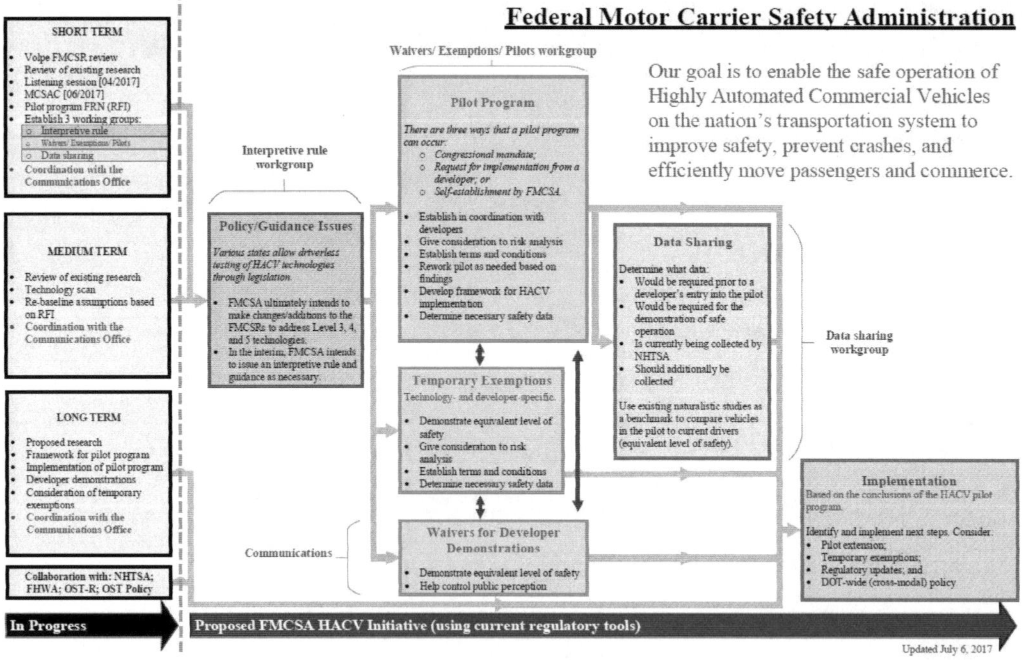

Considerations for Automated Truck Policy

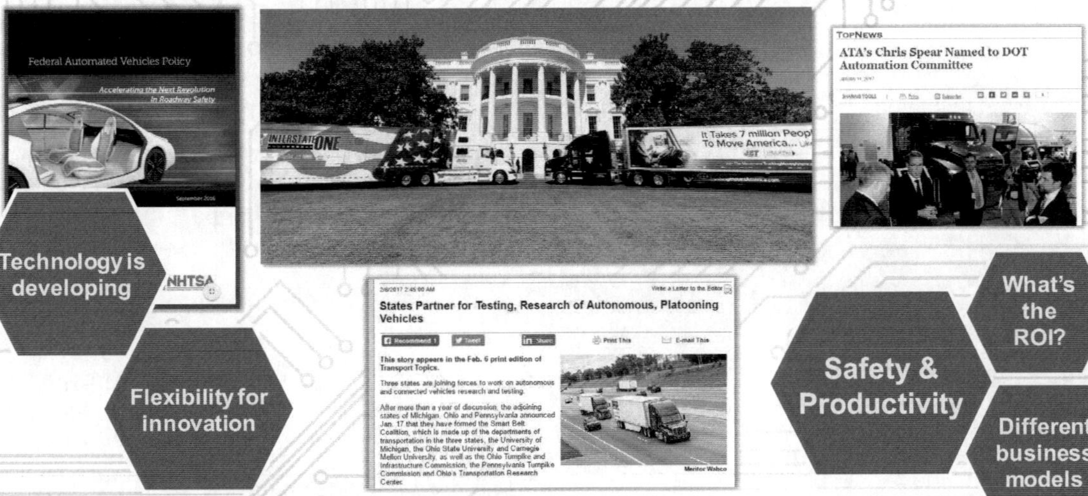

Technology is developing

Flexibility for innovation

Safety & Productivity

What's the ROI?

Different business models

Trucking.org

ATA

Connectivity and automated driving will allow for cleaner and safer megacities to grow and prosper

<u>There is no silver bullet</u>. Technology choices should be addressed by business needs and may vary.

Advanced
Technologies
Group
UBER

CDOT
CO
COLORADO
Department of Transportation
38th Ave
Park Ave
Downtown
EXIT
40
Identifying and Addressing Non-Technical Key Research Questions: Infrastructure
Shailen P. Bhatt, Executive Director
July 12, 2017

FY 2017-2018
$1.42 Billion Budget
ROADX
ACCELERATING TECHNOLOGY
CDOT RESPONSIBILITIES
ADMINISTERS $208 MILLION EACH YEAR IN FEDERAL GRANTS
3,454 BRIDGES
CDOT MAINTAINS & OPERATES 23,000 TOTAL LANE MILES OF HIGHWAY
DIVISION OF TRANSIT AND RAIL
ADMINISTERS FED/STATE GRANTS AND OPERATES BUSTANG
6.1 MILLION MILES PLOWED OF SNOW PER YEAR
35 MOUNTAIN PASSES OPEN YEAR-ROUND
AIRPORT PLANNING INTERFACE WITH FAA
Source: Colorado Department of Transportation, 2014

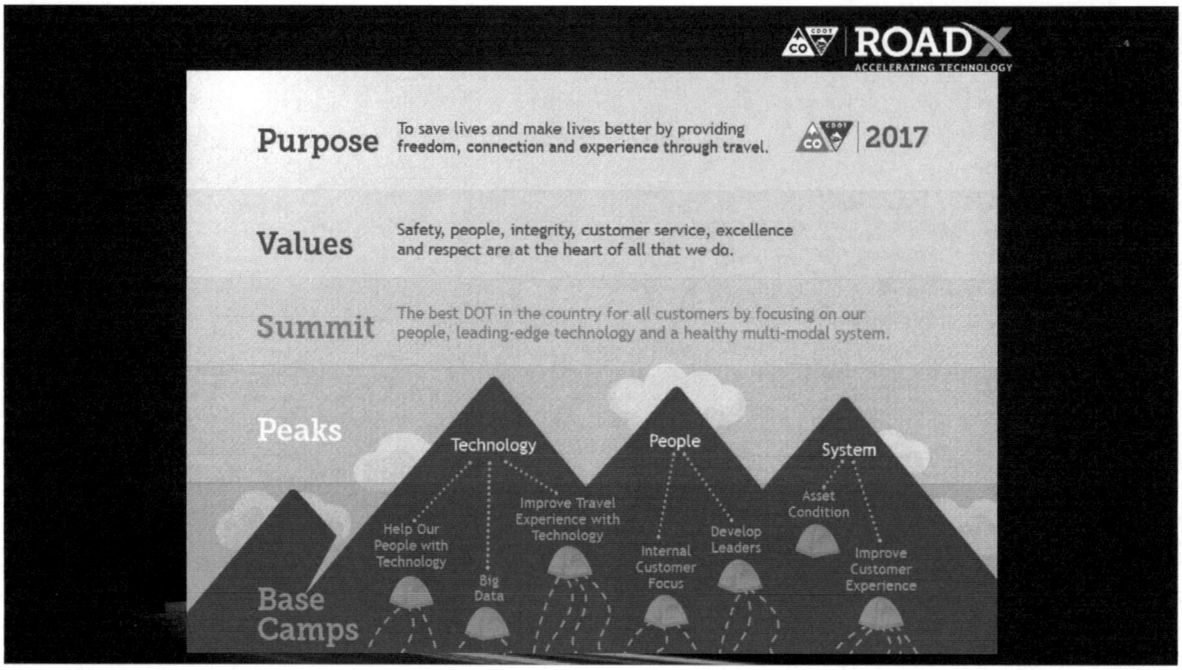
ROADX
ACCELERATING TECHNOLOGY
Purpose
To save lives and make lives better by providing freedom, connection and experience through travel.
2017
Values
Safety, people, integrity, customer service, excellence and respect are at the heart of all that we do.
Summit
The best DOT in the country for all customers by focusing on our people, leading-edge technology and a healthy multi-modal system.
Peaks
Technology
People
System
Help Our People with Technology
Big Data
Improve Travel Experience with Technology
Internal Customer Focus
Develop Leaders
Asset Condition
Improve Customer Experience
Base Camps

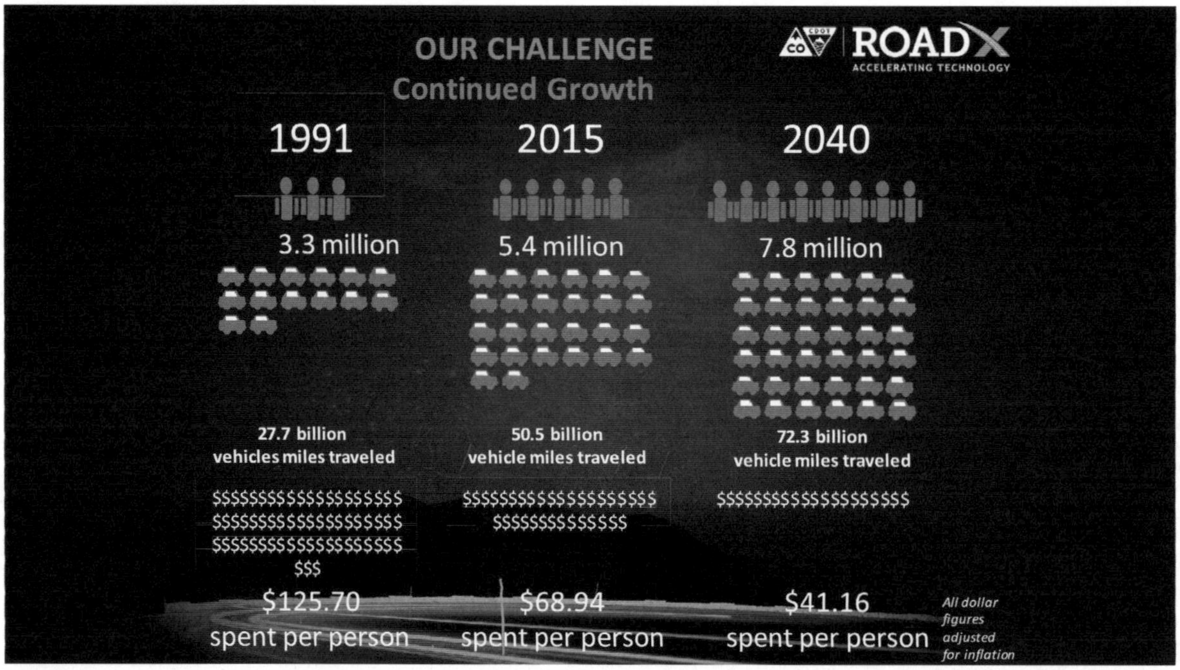
OUR CHALLENGE
Continued Growth
ROADX
ACCELERATING TECHNOLOGY
1991
2015
2040
3.3 million
5.4 million
7.8 million
27.7 billion vehicles miles traveled
50.5 billion vehicle miles traveled
72.3 billion vehicle miles traveled
$125.70 spent per person
$68.94 spent per person
$41.16 spent per person
All dollar figures adjusted for inflation

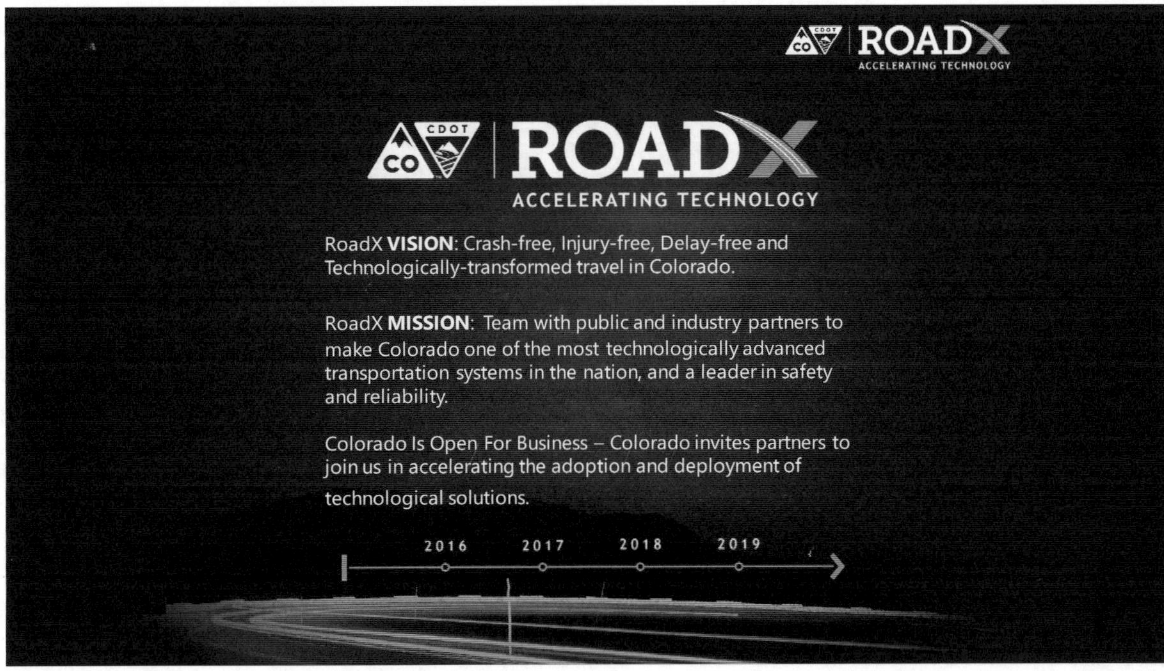
ROADX
ACCELERATING TECHNOLOGY
CDOT CO
ROADX
ACCELERATING TECHNOLOGY
RoadX VISION: Crash-free, Injury-free, Delay-free and Technologically-transformed travel in Colorado.
RoadX MISSION: Team with public and industry partners to make Colorado one of the most technologically advanced transportation systems in the nation, and a leader in safety and reliability.
Colorado Is Open For Business – Colorado invites partners to join us in accelerating the adoption and deployment of technological solutions.
2016
2017
2018
2019

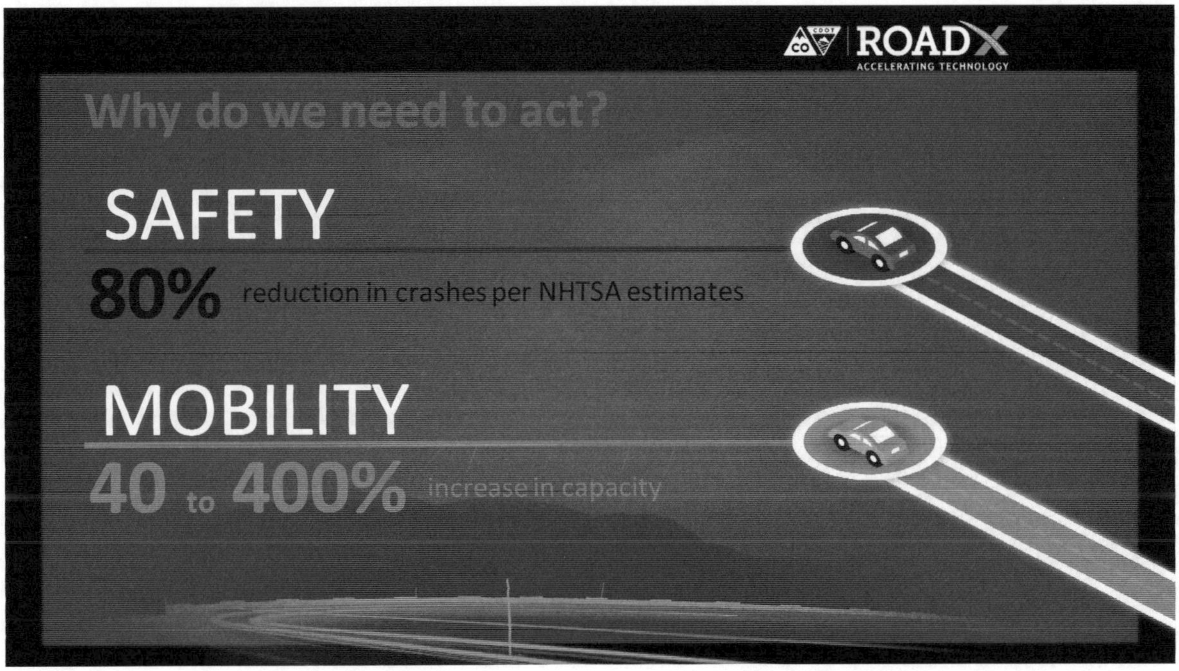
ROADX
ACCELERATING TECHNOLOGY
Why do we need to act?
SAFETY
80% reduction in crashes per NHTSA estimates
MOBILITY
40 to 400% increase in capacity

ROADX ACCELERATING TECHNOLOGY

5 levels of driving automation

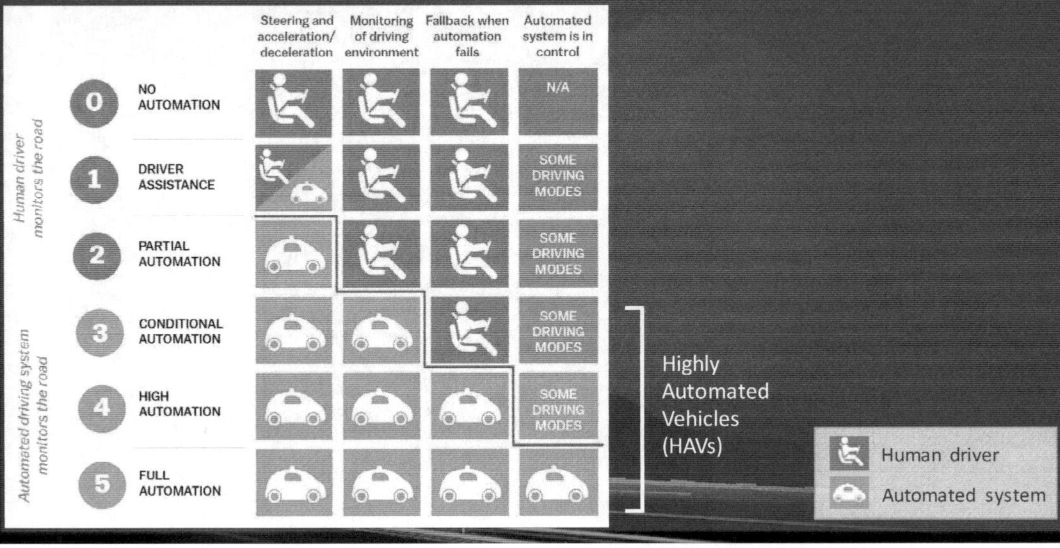

ROADX ACCELERATING TECHNOLOGY

NHTSA's AV Guidance and ODD

The document identifies **Operational Design Domain (ODD)** as the critical definition of where (such as what roadway types, roadway speeds, etc.) and when (under what conditions, such as day/night, normal or work zone, etc.) an HAV is designed to operate. The importance of communicating the ODD of an HAV to the consumer as part of broader product education is highlighted.

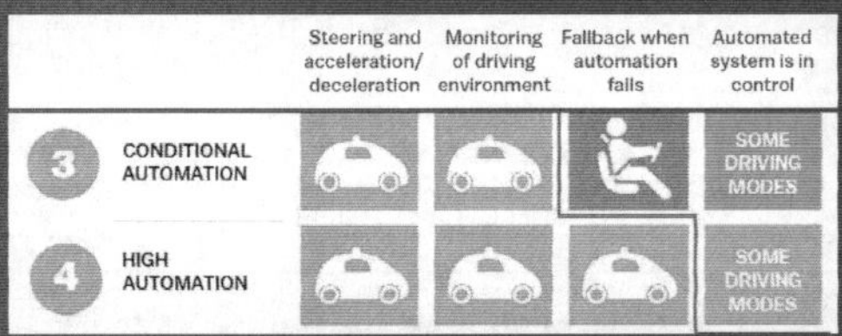

ROADX
ACCELERATING TECHNOLOGY

Connected road classification system

Level 1 Unpaved and/or non-striped roads designed to a minimum level of standard of safety and mobility

Level 2 Paved roads designed to AASHTO's standards with MUTCD signage. There is not Intelligent Transportation System (ITS) equipment or infrastructure to collect connected vehicle data (Dedicated Short Range Radio). Access to cellular date service may be available

Level 3 There is Intelligent Transportation System (ITS) equipment operated by a Traffic Operation Center (TOC) and/or, one way electronic data share between DOT/Vehicle/User and/or, mixed use lanes

ROADX
ACCELERATING TECHNOLOGY

Connected road classification system

Level 4 Roadway or specific lane(s) has adaptive ITS equipment (i.e. smart signals hold for vehicles, highway lighting that turn on for vehicles, etc.) with Traffic Operations Center override only, and/or two way data share between DOT/Vehicle/User, and/or lanes designated for vehicle levels 3 & 4 only

Level 5 (Advance Guide-way System) roadway or specific lane(s) designed for vehicle level 4 only with additional features that may include inductive charging, advance/enhanced data sharing, etc. Additionally, no roadside signs are needed as all roadway information is direct to vehicles' on-board systems

Level 6 All roadway elements designed for only vehicle level 5 systems – no signs, signals, striping... needed

Could solar roadways power our future?

COLORADO
Department of Transportation

ROADX
ACCELERATING TECHNOLOGY

Smart Highways

Smart 25
Ridegate to University

Software and sensor upgrades to better manage flow resulting in:

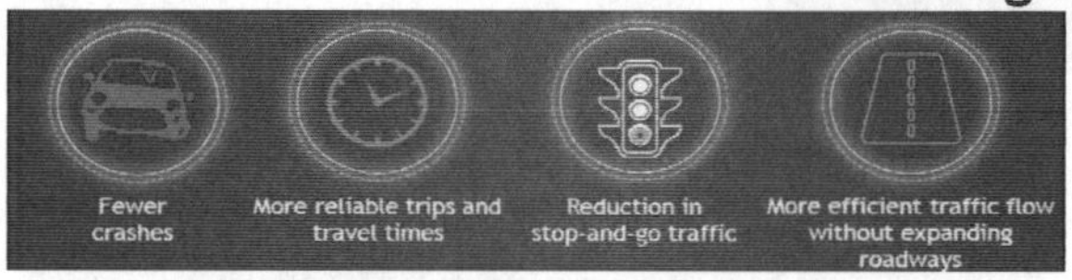

Fall 2017

here

Smart 70
Golden to Vail

Partnership with Here will provide real time data about hazards such as:

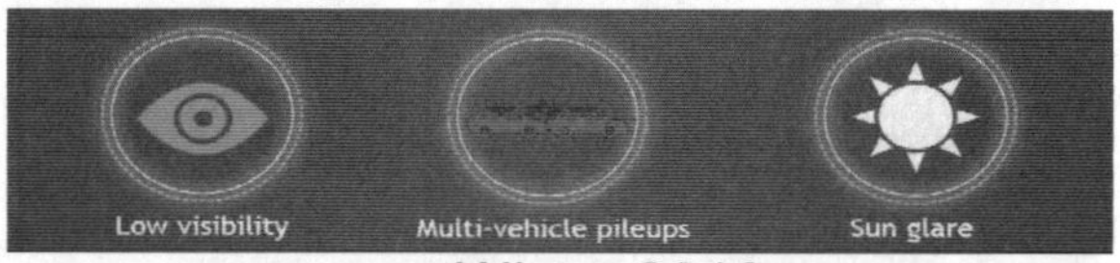

Winter 2016

Panasonic

Smart 70
Panasonic

Self-driving vehicles and infrastructure share data to eventually:

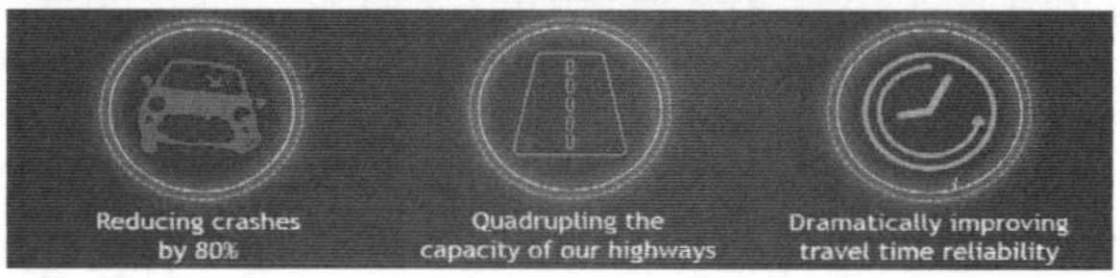

Ongoing

11

COLORADO
Department of Transportation

ROADX
ACCELERATING TECHNOLOGY

Innovation

OTTO **Transport**

Self driving vehicles, platooning, smart parking

Hyperloop

New way to move people and freight

Smart Powered Lane

Pilot to embed charging in roads to power electric vehicles

COLORADO
Department of
Transportation

Questions?

SERVING THE NEEDS OF ALL THROUGH BETTER DESIGN

Edward Steinfeld, ArchD
Director, IDeA Center

Automated Vehicles Symposium
July 11, 2017

University at Buffalo
Center for Inclusive Design and Environmental Access
School of Architecture and Planning

Center for Inclusive Design and Environmental Access (Mission)

- Founded in 1984, the primary goal of the IDeA Center is to produce knowledge and tools that will increase quality of life for groups who have been marginalized by traditional design practices.

- The IDeA Center is a multi-disciplinary research center with architects, urban planners, occupational therapists, industrial engineers, and social scientists.

University at Buffalo
Center for Inclusive Design and Environmental Access
School of Architecture and Planning

Our Work

- We are the leading R&D center on universal design and environmental accessibility in North America
- Laboratory simulations to test out new features and products
- Field research to study systems in use
- Surveys of stakeholders
- Publications, including the first comprehensive text on universal design and the first book on accessible public transportation (upcoming)
- Continuing education

University at Buffalo
Center for Inclusive Design and Environmental Access
School of Architecture and Planning

Universal Design

A *process* that enables and empowers a diverse population by improving human performance, health and wellness, and social participation.

Steinfeld and Maisel (2012)

Elevator message:

Making things easier to use, healthier, and friendlier.

University at Buffalo
Center for Inclusive Design and Environmental Access
School of Architecture and Planning

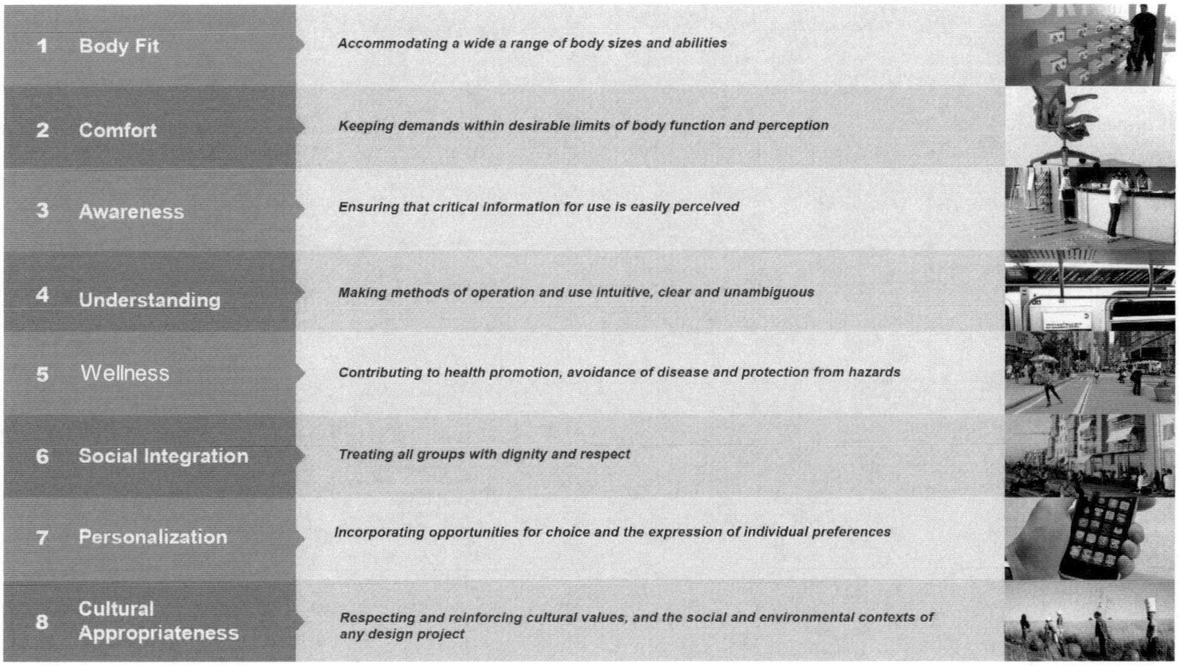
1 Body Fit
Accommodating a wide a range of body sizes and abilities
2 Comfort
Keeping demands within desirable limits of body function and perception
3 Awareness
Ensuring that critical information for use is easily perceived
4 Understanding
Making methods of operation and use intuitive, clear and unambiguous
5 Wellness
Contributing to health promotion, avoidance of disease and protection from hazards
6 Social Integration
Treating all groups with dignity and respect
7 Personalization
Incorporating opportunities for choice and the expression of individual preferences
8 Cultural Appropriateness
Respecting and reinforcing cultural values, and the social and environmental contexts of any design project

Unleashing Potential
Google
Photo credit: Screenshots Self-Driving Car Test: Steve Mahan

The Broader Market

Reducing Dependency
IT 316-2153
2806
Photo credit: http://www.impruve.org/

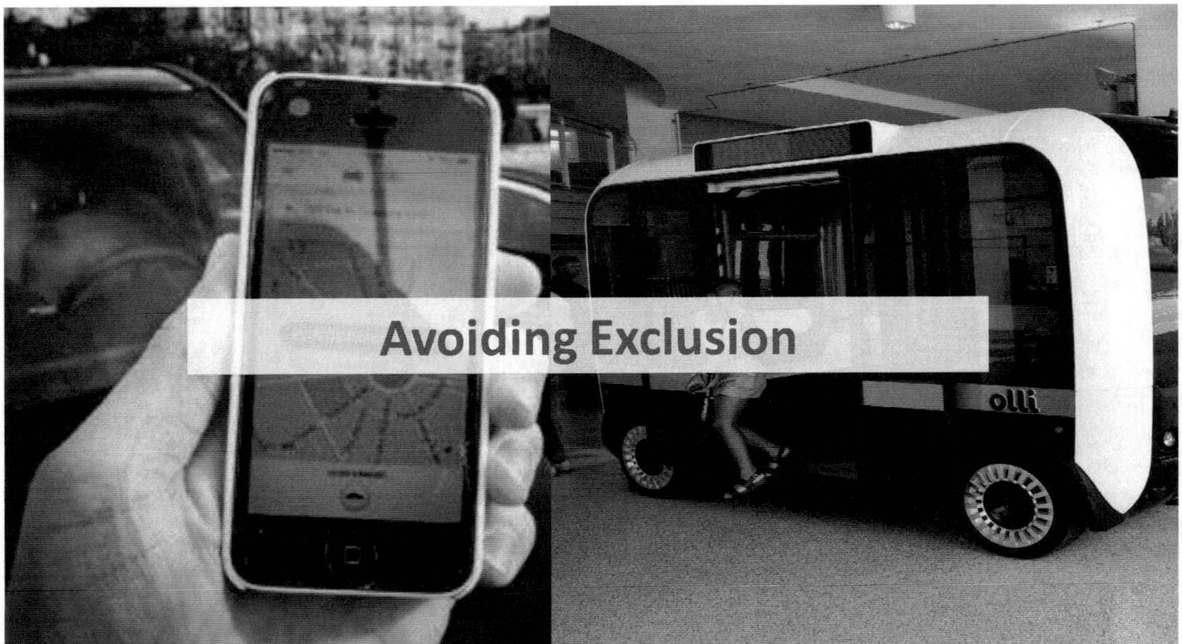
Avoiding Exclusion
olli
Photo credit: https://phys.org/news/2016-09-uber-groundbreaking-driverless-car.html
Photo credit: http://washington.cbslocal.com/2016/06/18/driverless-bus-to-begin-frequenting-d-c-streets

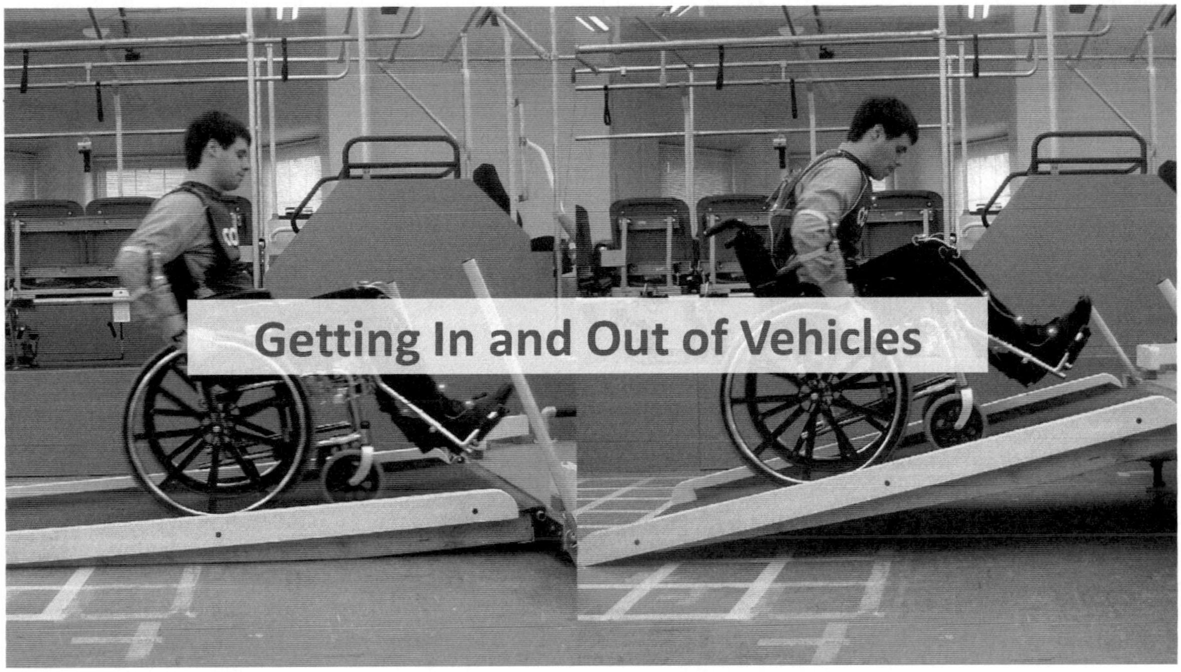
Getting In and Out of Vehicles

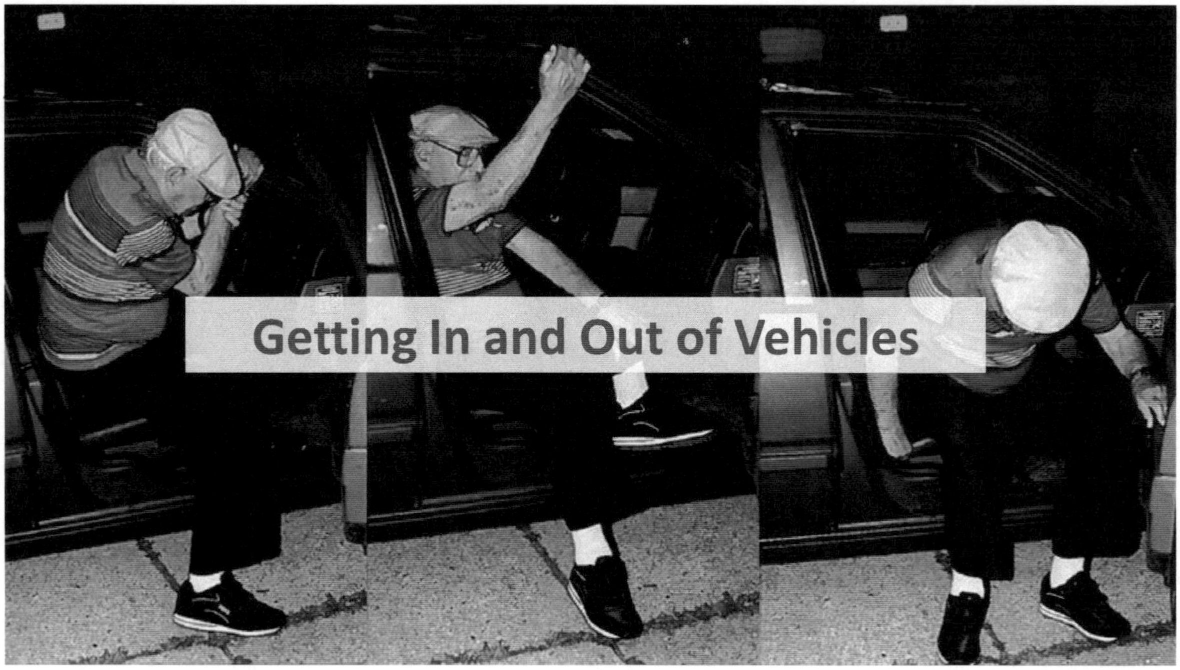

Getting In and Out of Vehicles

Problems reported by 100 older drivers and passengers (Steinfeld et al. 1999)

Getting legs in and out	Bending head down to get in the door
Fastening seat belts	Storing mobility devices
Sliding in and out	Bending at the waist while getting in or out
Rear seat access	Opening Doors
Step down into the vehicle	Storing packages
Stepping up into the vehicle	Using the door locks
Reaching doors to close	Using door handles

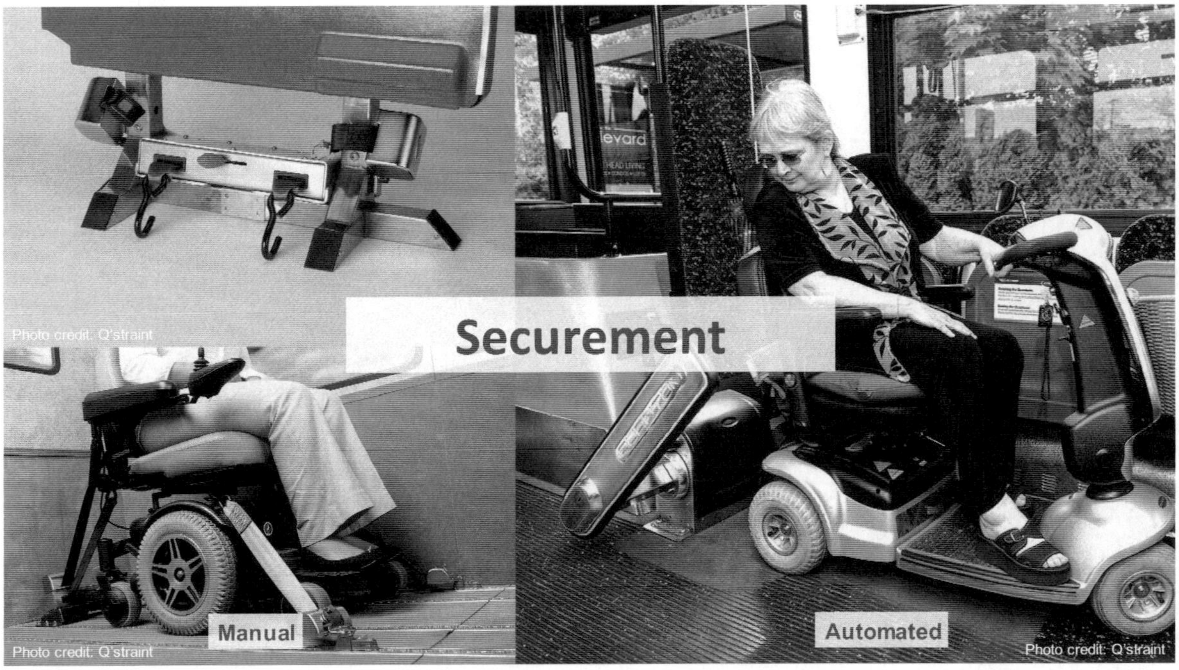

Securement

4–Point (n=36)

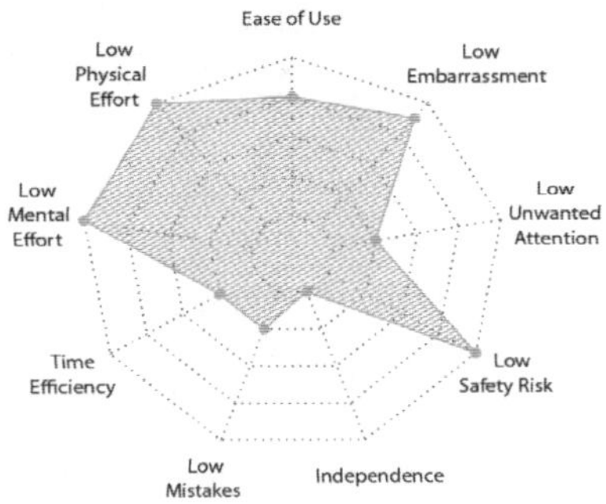

Quantum (n=36)

University at Buffalo
Center for Inclusive Design and Environmental Access
School of Architecture and Planning

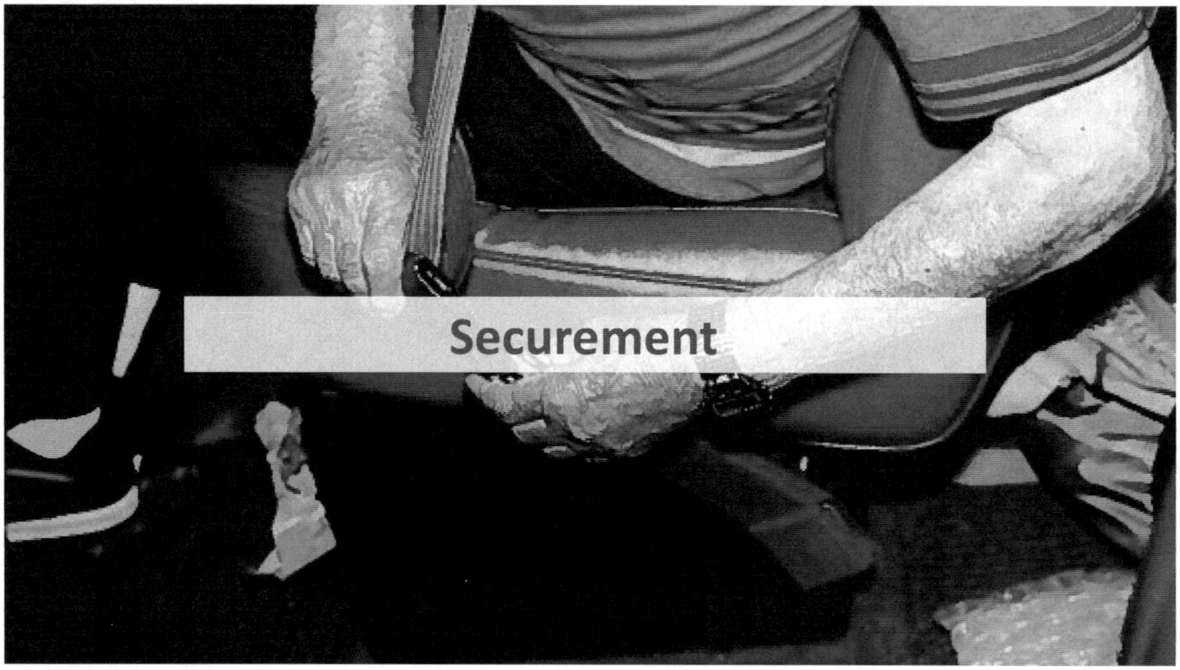
Securement

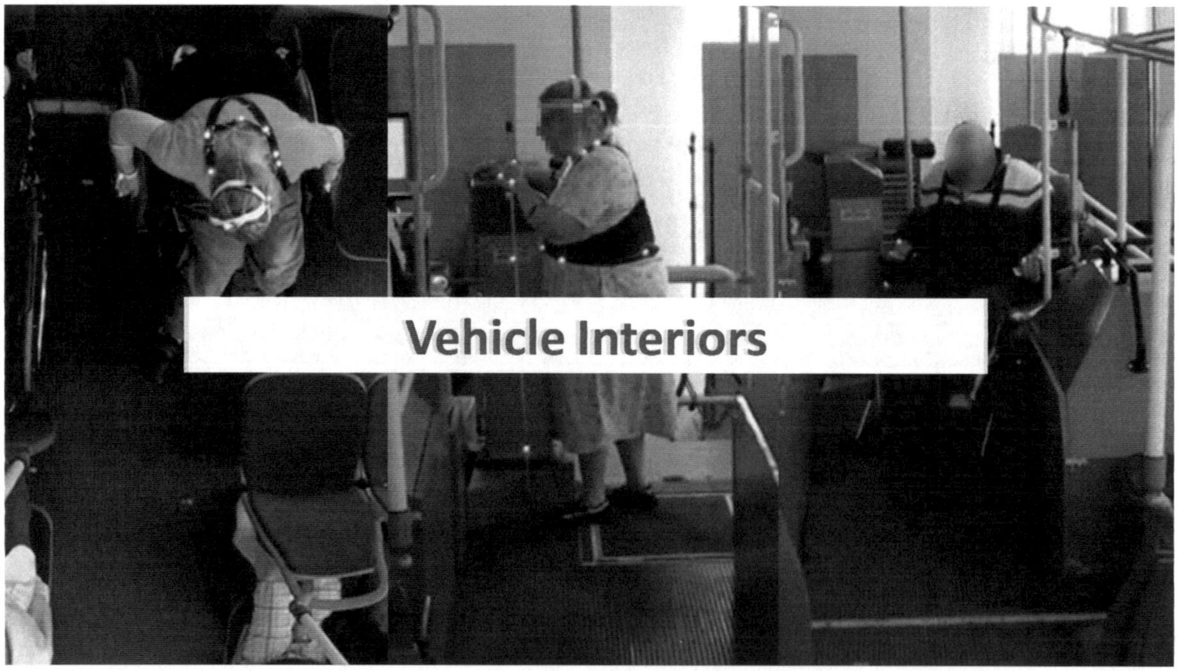
Vehicle Interiors

Communications

University at Buffalo
Center for Inclusive Design and Environmental Access
School of Architecture and Planning

Concluding Messages

- Practicing universal design brings the benefits of automated vehicles to all
- Consider the diversity of riders at the conceptual stage and follow through with constant improvement
- Get out ahead of regulators by adopting an inclusive approach
- Use universal design to improve performance and broaden markets

University at Buffalo
Center for Inclusive Design and Environmental Access
School of Architecture and Planning

IDeA

Center for Inclusive Design & Environmental Access
School of Architecture & Planning
State University of New York at Buffalo

tel: +1 (716) 829.5899
fax: +1 (716) 829.3861
email: arced@buffalo.edu
web: http://idea.ap.buffalo.edu
facebook: http:/www.facebook.com/ideacenterUB/
twitter: @IDeA_Center

University at Buffalo
Center for Inclusive Design and Environmental Access
School of Architecture and Planning

An Unexpected Uptick?

In an automotive ecosystem with rapidly increasing automation (assistance and mitigation systems), why aren't the roads safer?

- Preliminary estimates from the National Safety Council indicate 40,200 motor-vehicle deaths on US roads in 2016
  - 6% increase over 2015
  - 14% increase over 2014
  - Highest in 9 years
- Motor-vehicle mileage up 3% in 2016

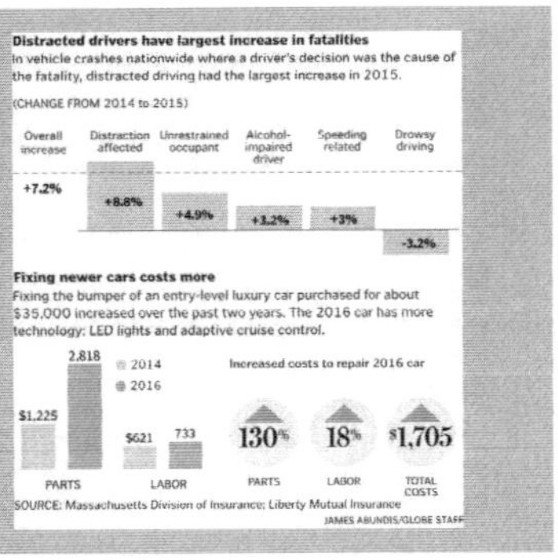

Data and Figure drawn from the National Safety Council 2016, 2017 and The Boston Globe 2017

2

Automation as an Evolving Process

We are talking about it more but the ship set sail decades ago!

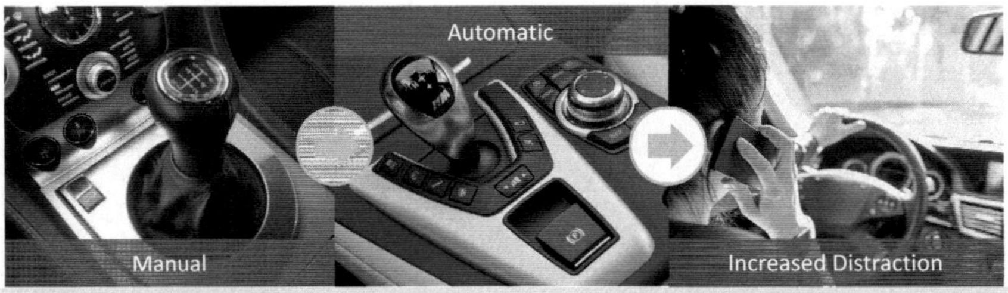

Automatic transmissions reduce drivers' operational workload.
Drivers have used the "freed up" resources to do other things!

The Advanced Vehicle Technology Consortium

- **Originators:** MIT AgeLab, Touchstone Evaluations & Agero
- **Founding Members:** Delphi, Liberty Mutual, Jaguar Land Rover, Autoliv, Toyota
- **Full Members:** TBD
- **Affiliate Members:** Consumer Reports & TBD
- **Focus:** To collect cutting edge data that objectively characterizes the behavioral and safety benefit of advanced driver assistance systems, higher levels of automation, and other in-vehicle technologies under real-use conditions
- **Outcomes:** A data driven understanding of drivers' utilization of advanced vehicle technologies & the development of data and analytic methods necessary for answering important member common questions about AVT

Investigating Automated Technology Use in the Wild

MIT AVT
Naturalistic Data

Study months to-date: **18**
Participant days: **4,929**
Drivers: **60**
Vehicles: **24**
Trips taken: **27,843**
Hours driven: **7,565**
Miles driven: **138,188**
Video frame: **2.6 billion**
Video pixels: **2.4 quadrillion**

Study data collection is ongoing.
Statistics updated on: July 2, 2017

MIT AGELAB | MIT

Technologies are Improving Post-Production

Improved Volvo Pilot Assist performance with a dealer delivered upgrade.

MIT AGELAB | MIT

Use Outside of the Operational Design Domain, Is It Foreseeable Misuse?

It is difficult to design and test for all use conditions; is a learning approach or geofencing critical?

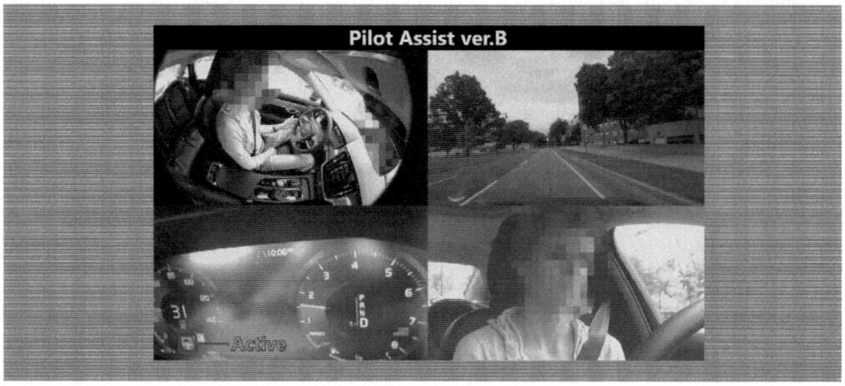

7

The Perils of Misuse Outside of the Operational Design Domain

How do experiences impact willingness to use, trust, safety?

8

Insight on Tesla Autopilot Use

16,422 transfers of control across 454 hours of Autopilot use

- Human to machine
  - From available – 7,308
  - From ACC – 903
- Machine to human (human initiated)
  - To manual – 6,911
  - To ACC – 1,300
- Machine to human (system initiated)
  - AI assistance required – 17
  - ODD speed boundary (90 MPH) exceeded – 12
  - Intersection (slow speed car following through turn) – 5
  - Automated lane change (e.g. lane characteristic changes) – 4
  - Other - 4

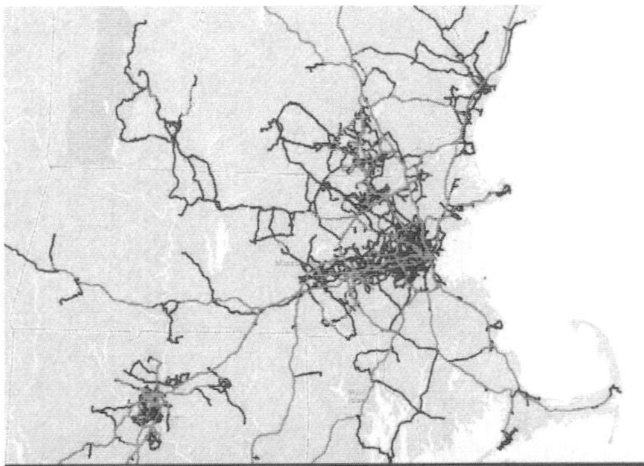

GPS points with Autopilot engaged (blue) overlaid on manual control

Linking Theory to Actual Behavior

When has a driver accepted a transition? regained full awareness?
Further, drivers demonstrate different use styles.

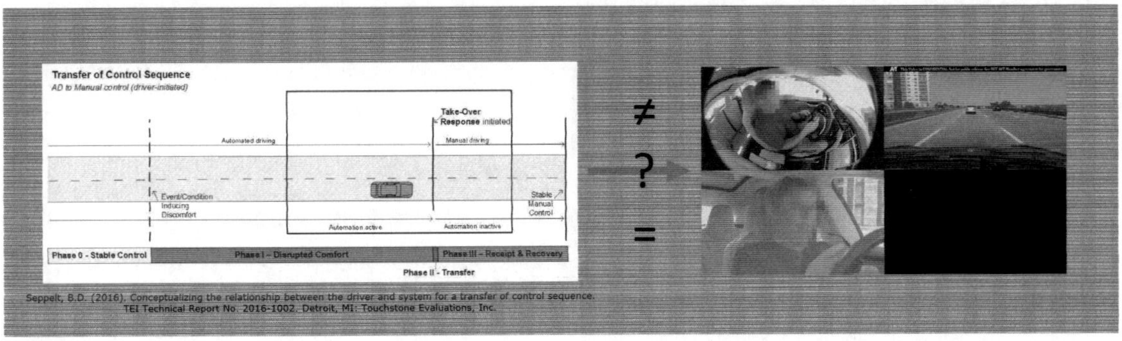

Using Naturalistic Data to Teach Robots to Drive

Approach: End-to-end deep neural networks for perception and steering control

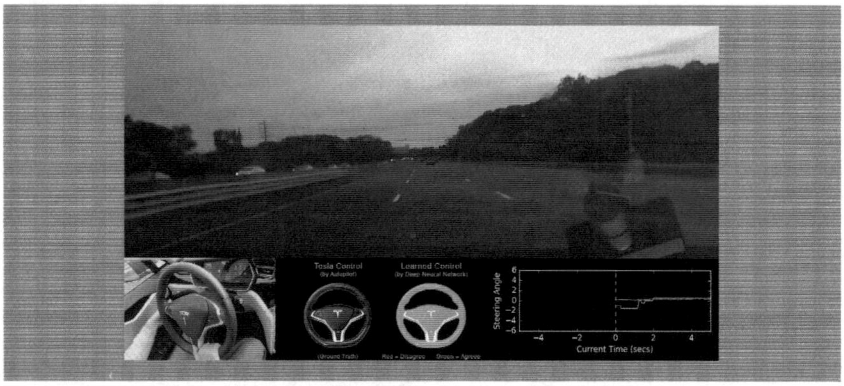

MIT 6.S094: Deep Learning for Self-Driving Cars - http://selfdrivingcars.mit.edu/

Arguing Machines

Goal: Predicting the resumption of control from Autopilot.

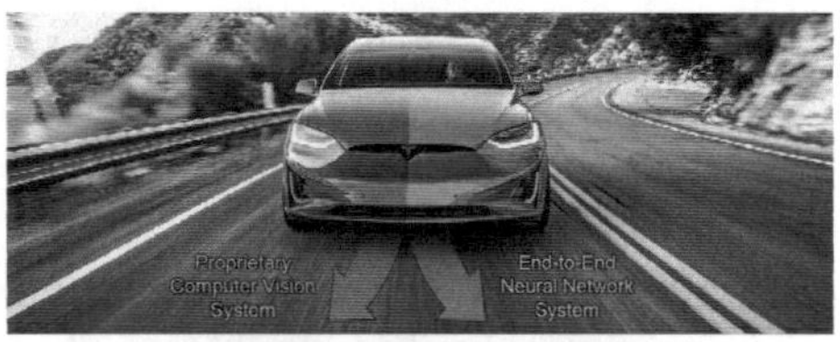

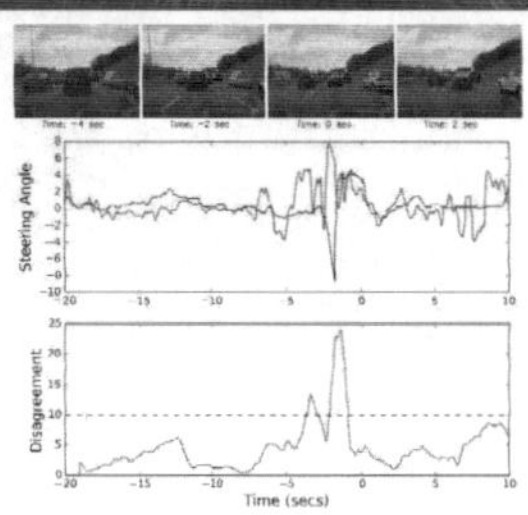

90% accuracy in a 6 sec window

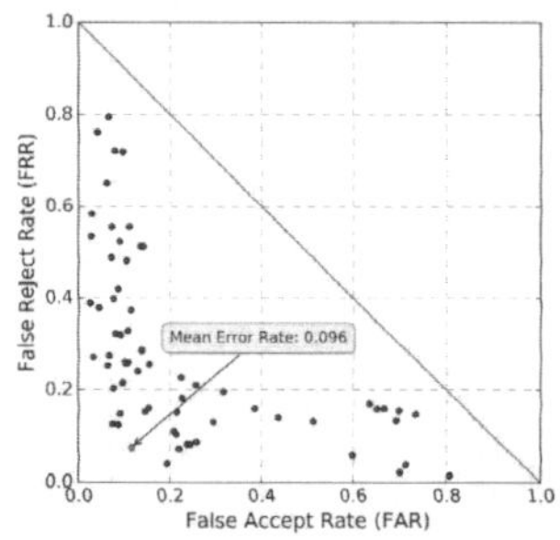

The Future May Be One of Relatively "Novice" Drivers

Vehicle Miles Traveled (VMT) **Vehicle Miles Driven (VMD)**

Is Driving a Privilege or a Right?

It is plausible some drivers will exploit the bounds of risk under many (if not all) driving contexts

The challenge is engineering for all!
Do we now need to be designing to the outliers?

Many Factors will Influence the Speed of Deployment

Human centered technologies and proactive policy will play a key role

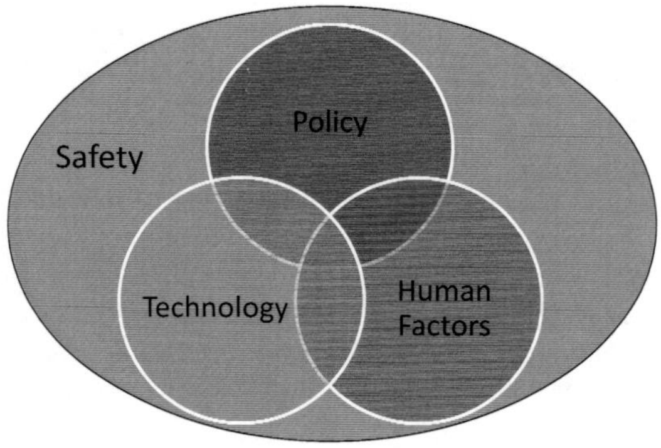

The Future May Be Autonomous, But…

Humans will continue to have a role for some time

- Level II systems introduce a number of use challenges
- Reported experiences combined with quantitative data tell a story
  - Mode confusion
  - Role confusion
  - Misplaced trust
- Investment in human centered engineering throughout the automotive ecosystem is critical to a safe mobility future
  - Improved communication with the driver
  - Fused decision models that consider driver state
  - Consumer education and training (dealer delivery and coaching)

AUTOMATED VEHICLES SYMPOSIUM 2017

USERS. VEHICLES. INFRASTRUCTURE.

July 12, 2017

QUANERGY

LiDAR and the Autonomous Vehicle Revolution for Truck and Ride Sharing Fleets

Louay Eldada

CEO and Co-founder, Quanergy Systems

QUANERGY

New Paradigm in 3D Sensing

- Disruptive Technologies:
  - Solid State 3D LiDAR sensors
  - Advanced embedded processors
  - Inertial Measurement Units (IMU)

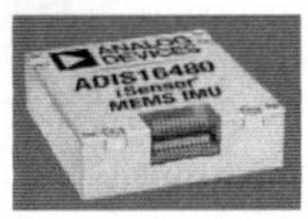

- Advanced Systems:
  - ADAS, Autonomous Vehicles
  - Smart Homes, Smart Security
  - Robots, Drones, 3D-Aware Devices

- Smart Solutions:
  - Global daily-updated cm-accurate 3D HD Map
  - GPS-free Navigation through SLAM
  - Smart IoT

 2

QUANERGY

Our Mission

Social Impact

Save Lives

Save Energy

Save Space

Save Time

Save Money

Safety

Efficiency

QUANERGY

Our Position

- Quanergy is the leader in advanced sensing innovation
- Leadership position is in both hardware and software, enabling full system solutions

Awards & Recognition:

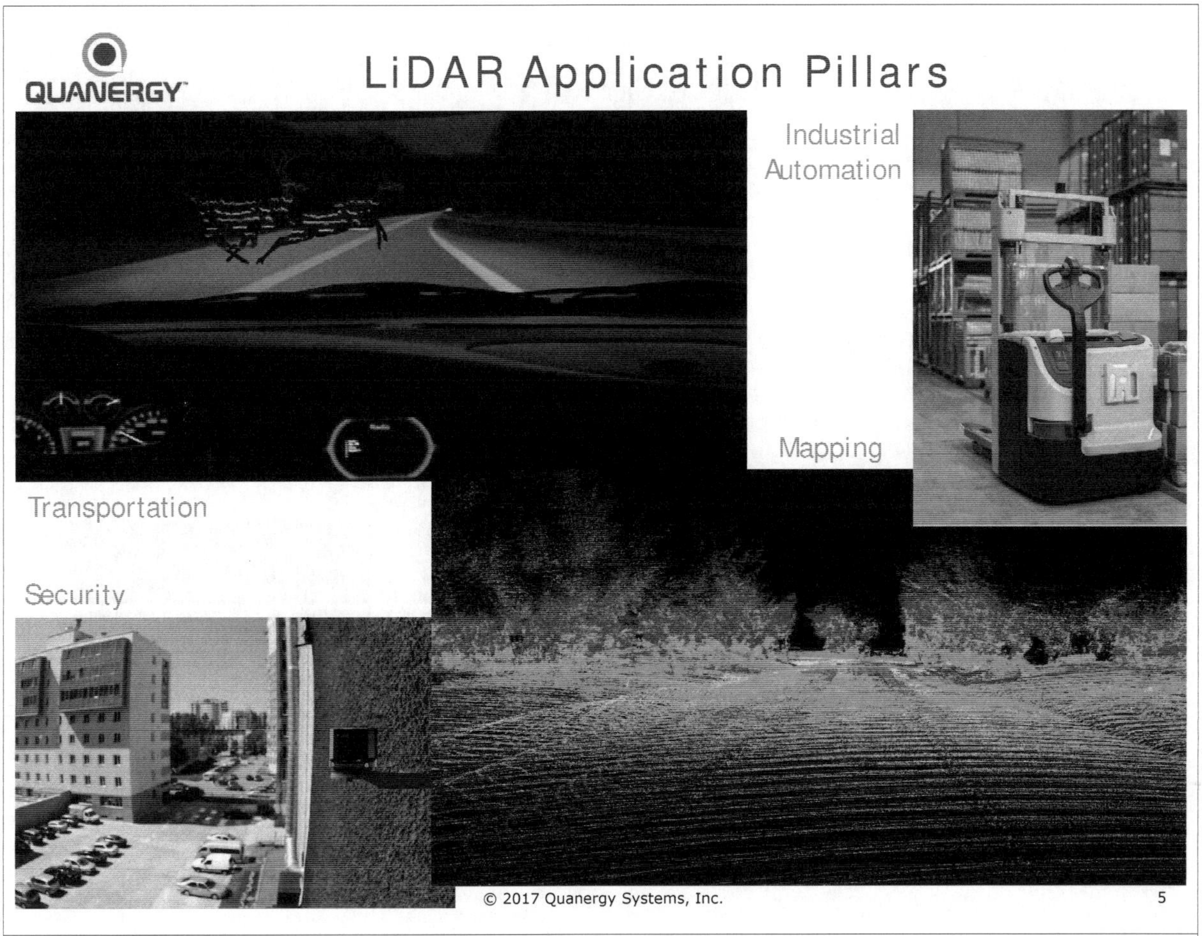
QUANERGY
LiDAR Application Pillars
Industrial Automation
Mapping
Transportation
Security
© 2017 Quanergy Systems, Inc.
5

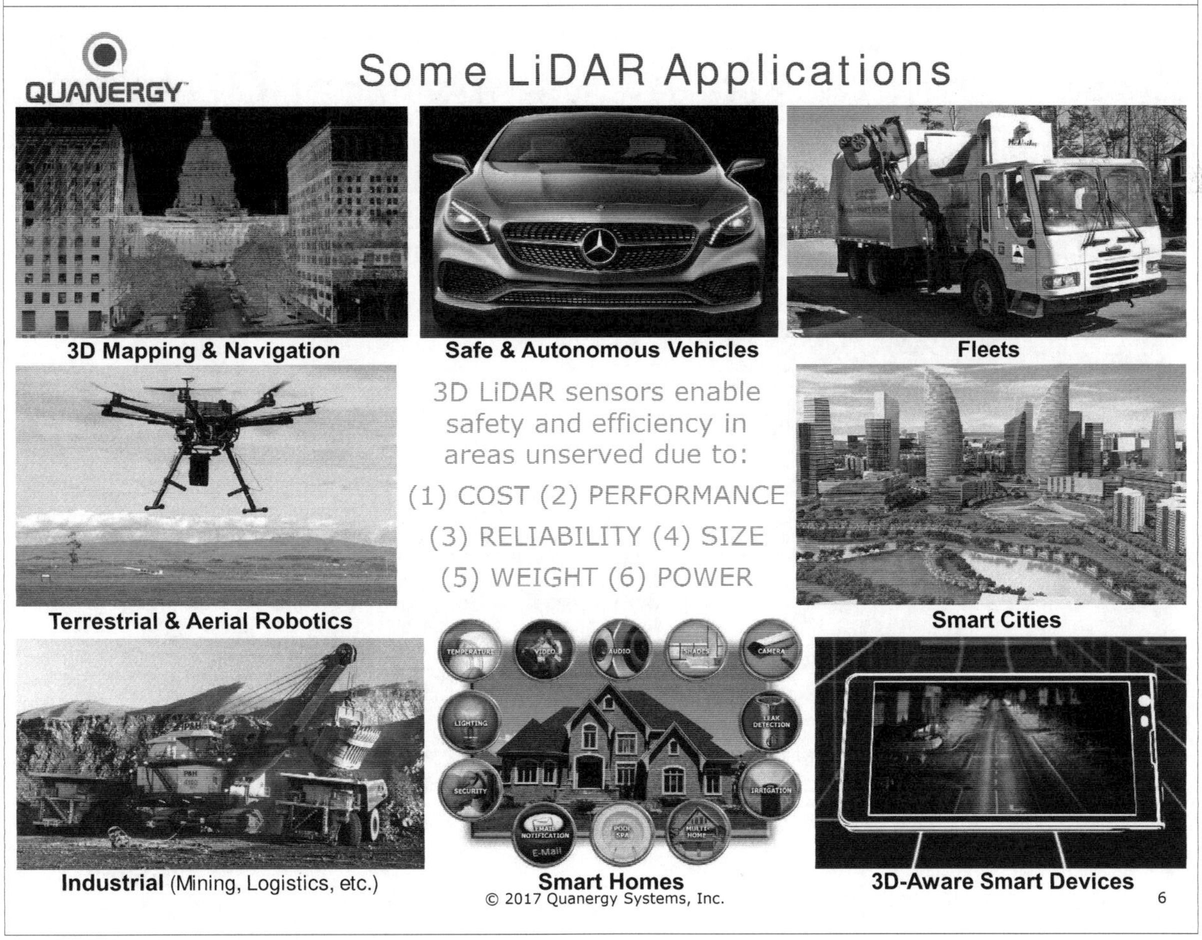
QUANERGY
Some LiDAR Applications
3D Mapping & Navigation
Safe & Autonomous Vehicles
Fleets
3D LiDAR sensors enable safety and efficiency in areas unserved due to:
(1) COST (2) PERFORMANCE
(3) RELIABILITY (4) SIZE
(5) WEIGHT (6) POWER
Terrestrial & Aerial Robotics
Smart Cities
Industrial (Mining, Logistics, etc.)
Smart Homes
3D-Aware Smart Devices
© 2017 Quanergy Systems, Inc.
6

Why LiDAR

LiDAR is the most accurate perception sensor:
- 3D shape with high resolution up to max range
- Distance with high accuracy
- Orientation
- Intensity

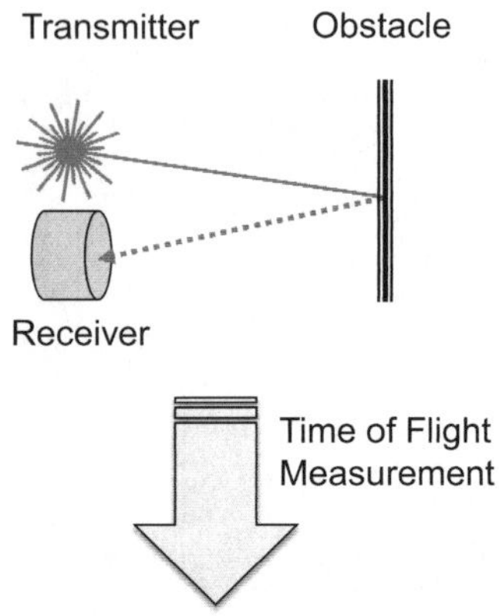

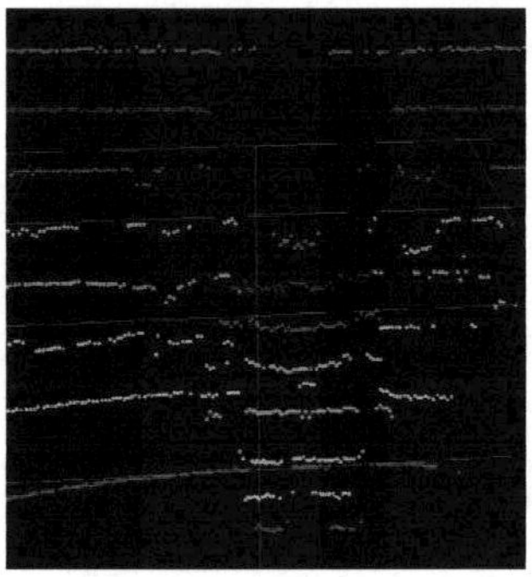

	LiDAR	Radar	Video
Sensing Dimensions	3D	1D	2D
Range	+++	+++	×
Range Rate	++	+++	×
Field of View	+++	++	+
Width & Height	+++	×	+
3D Shape	+++	×	×
Object Rec @ Long Range	+++	×	×
Accuracy	+++	×	+
Rain, Snow, Dust	++	+++	×
Fog	+	+++	×
Pitch Darkness	+++	+++	×
Bright Sunlight	+++	+++	×
Ambient Light Independence	+++	+++	×
Read Signs & See Color	+	×	+++

Solid State vs. Mechanical LiDAR

Mechanical LiDAR
Expensive, Large, Heavy, High Power,
Low Performance, Low Reliability

Solid State LiDAR (Quanergy S3)
Low Cost, Compact, Lightweight,
Low Power, High Performance,
High Reliability

QUANERGY

Product Roadmap

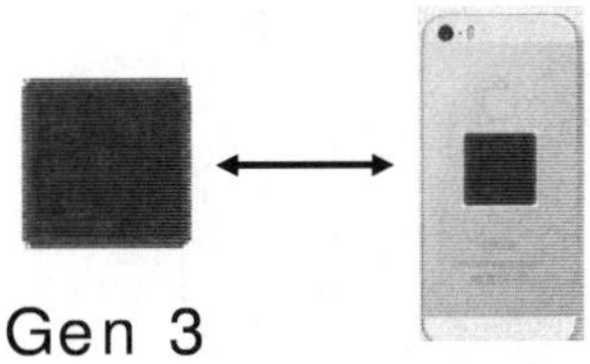

Gen 3
Solid State
(S3 ASIC)

Gen 2
Solid State
(S3, S3-Qi)

Gen 1
Mechanical
(M8)

- 1 Patent Granted
- 9 Patents Pending
- 15 Patent Apps in Prep

MCM: Multi Chip Module
ASIC: Application Specific Integrated Circuit 9

QUANERGY

S3 ICs

S3 LiDAR

Main S3 ICs, all based on **Si CMOS**:

- **Transmitter:**
  - OPA (Optical Phased Array) Photonic IC
  - OPA Control ASIC for beam forming/steering
- **Receiver:**
  - SPAD (Single Photon Avalanche Diode) Array IC
  - ROIC (Read Out IC) with TDC (time-to-digital converter) circuitry
- **Processor:**
  - FPGA for processing raw data into point cloud
  - ARM-based processor for data fusion, perception and object list generation

 10

QUANERGY

S3 Operation Principle

S3 Inside View

SPAD (Single Photon Avalanche Diode - Geiger mode APD) Array IC

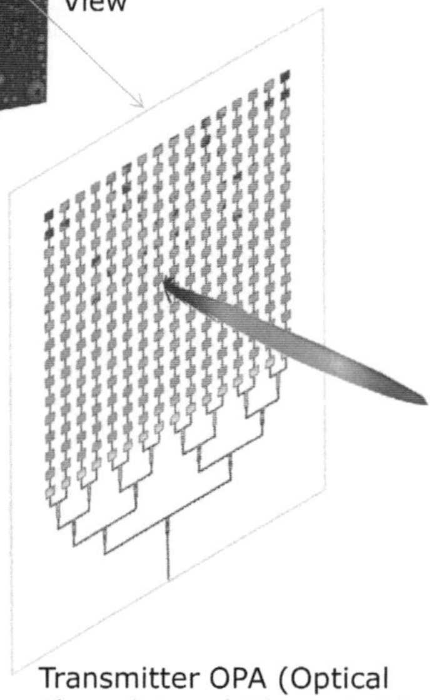

Transmitter OPA (Optical Phased Array) Photonic IC with far field radiation pattern (laser spot)

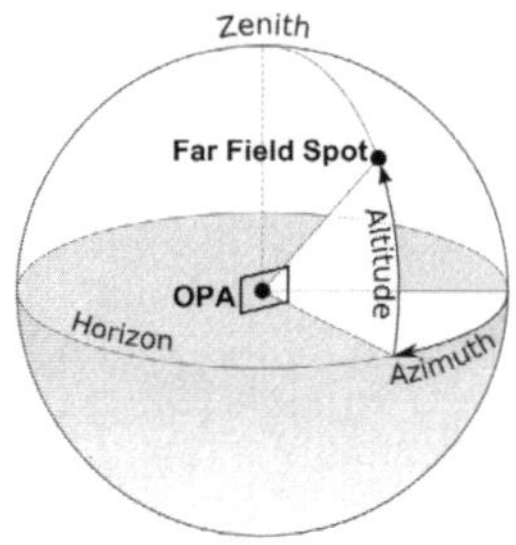

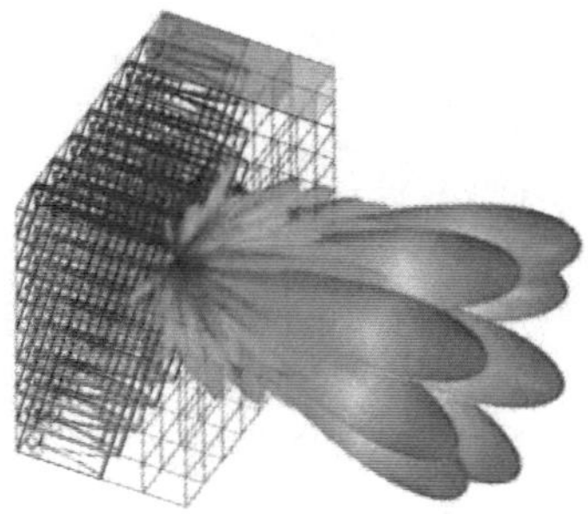

Overlaid far-field patterns for various steering angles

QUANERGY

S3 Performance & Unique Capabilities

- No mechanical moving part – no wear, no misalignment, no recalibration, no eye safety issue, MTBF > 100,000hrs operation
- Large FOV (120°), long range (150m)
- High resolution, high accuracy (at 100m spot size < 5cm, accuracy < 5mm)
- Software defined beam forming
  - Zoom in/zoom out for coarse & fine view
  - Random access in entire FOV, no sequential scanning required
  - Adjustable window within field of view
  - Arbitrary point density within each frame
  - Software defined frame rate

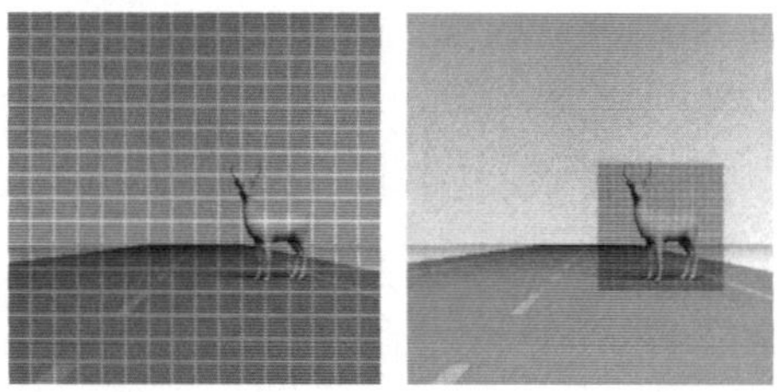

QUANERGY

Headlight-Integrated S3

- Developed with Koito, largest global maker of automotive headlights, first automotive headlight with built-in LiDAR sensors
- Each headlight, located on a corner of a vehicle, incorporates two S3 solid state LiDARs that sense forward and to the side
- The headlight protects the sensors from dust, dirt and water, and lens washers ensure an unobstructed view for the sensors
- World Premiered at CES 2017

QUANERGY

KOITO

13

QUANERGY

3D Composite Point Cloud
Terrestrial Mapping

14

3D Composite Point Cloud
Aerial Mapping

 15

3D Composite Point Cloud
Aerial Mapping

 16

Advanced Driver Assist Systems (ADAS)

Lane Keeping

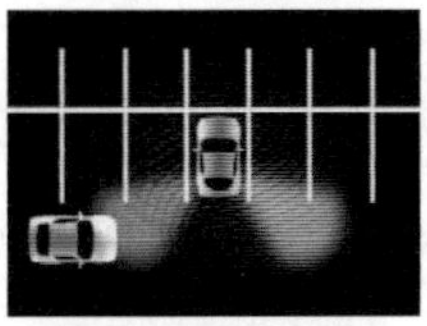
Parking Assist

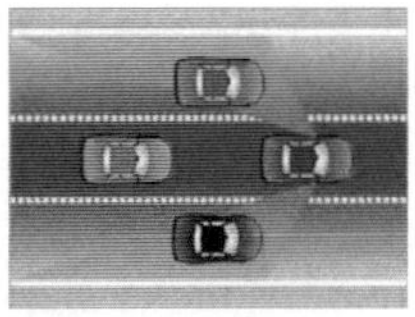
Blind Spot Detection

Adaptive Cruise Control
& Traffic Jam Assist

Front/Rear Collision Avoidance

Cross Traffic Alert &
Intersection Collision Avoidance

Autonomous Emergency Braking
& Emergency Steer Assist

Object Detection,
Tracking, Classification

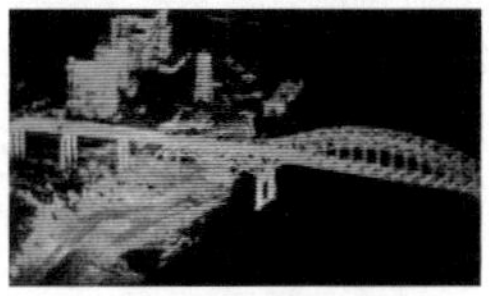
Scene Capture &
Accident Reconstruction

Typical Commercial ADAS

- **Video camera**
- **50° RADAR Sensor**
- **12° RADAR Sensor**
- **LiDAR Sensor**

Driver Blind Spot

Areas not blue/pink: System Blind Spot

Volvo System Built by Delphi

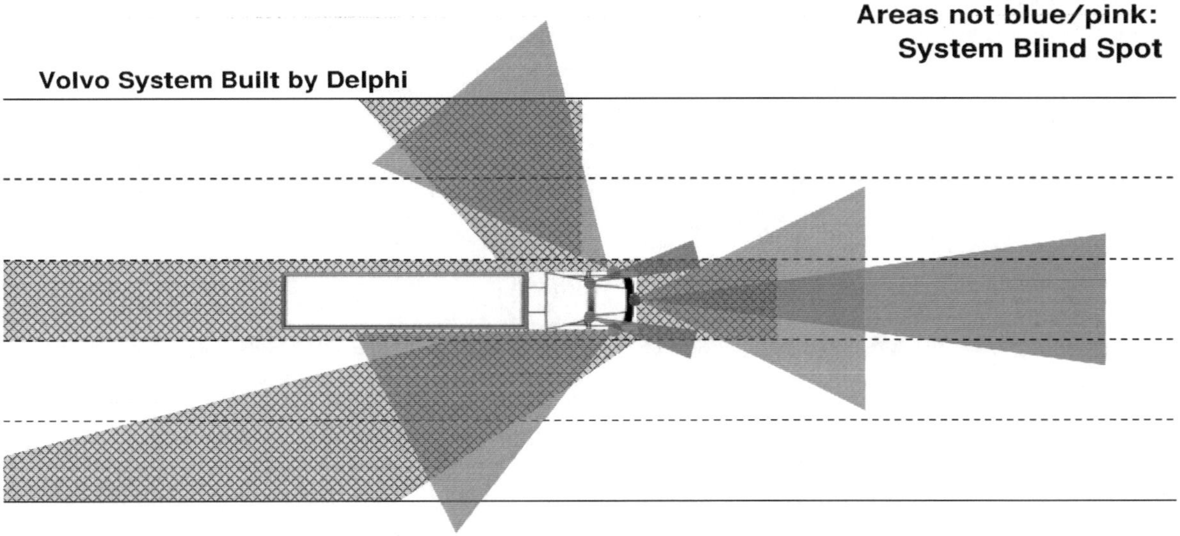

Quanergy ADAS

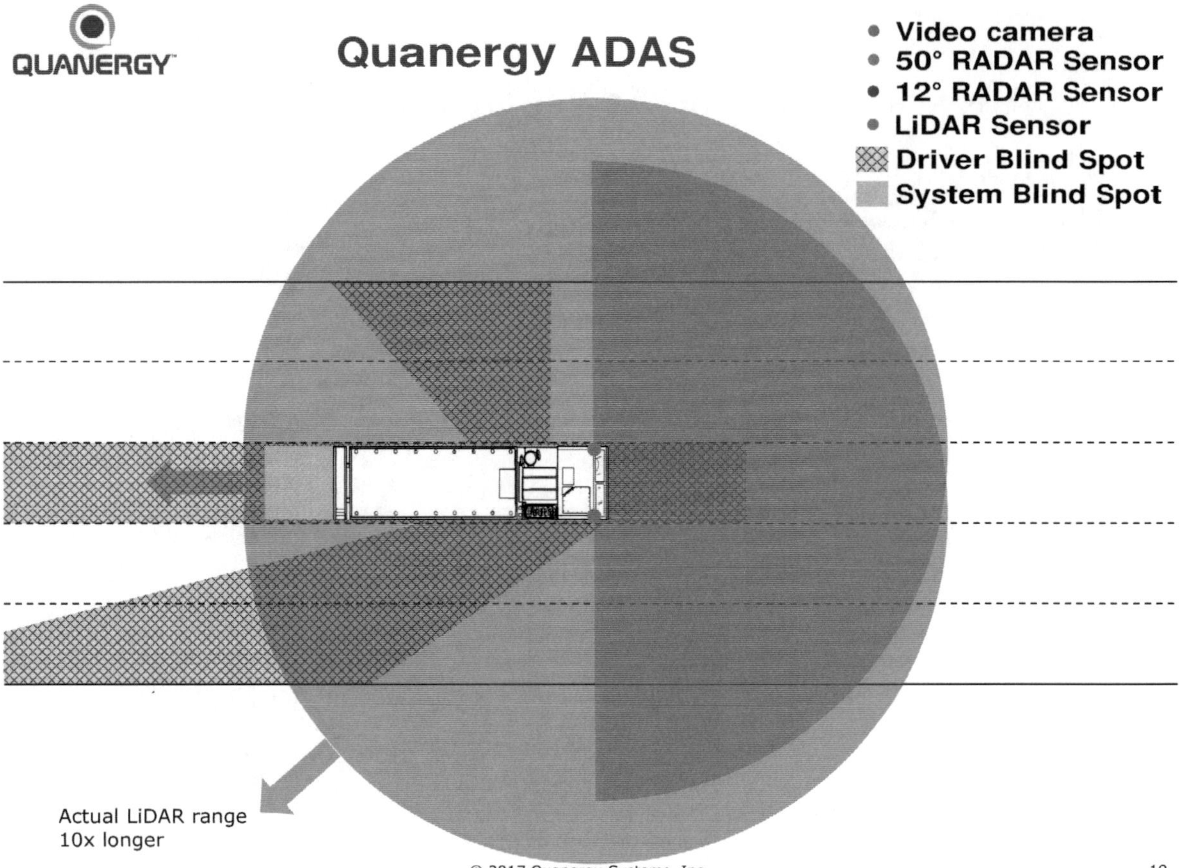

Autonomous Vehicle Sensors

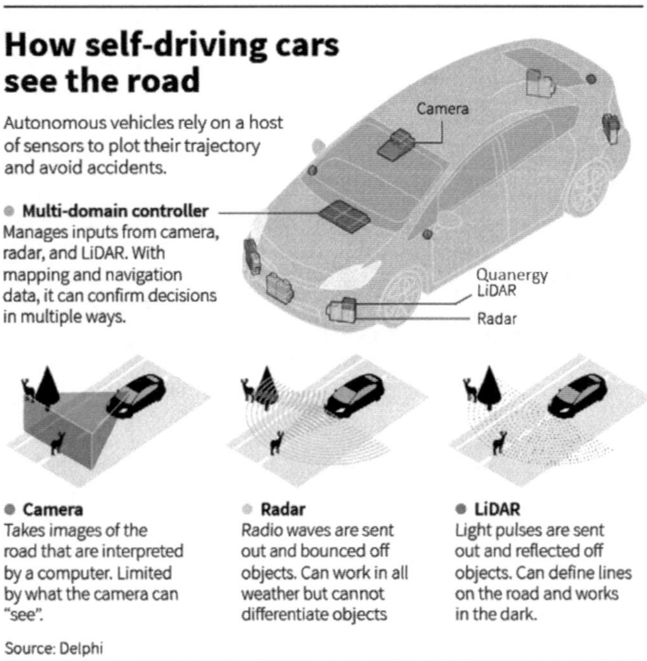

LiDAR is the primary sensor. As such, it is used for:

1. Perception
2. Localization
3. Navigation

Autonomous Driving Based on Deep Learning

Perception Pipeline

Partner NVIDIA

1. **Vehicle LiDAR Raw Input**
   Corrected point cloud using IMU, and video frames
2. **Occupancy Map**
   Created using LiDAR-video sensor fusion, probabilistic map informs which voxels are likely occupied
3. **Object Detection & Tracking**
   Run LiDAR and video output into a neural network that was trained to recognize and classify objects (cars, bikes, pedestrians, etc.)
4. **Localization**
   Determine position by registering within a pre-existing HD map, localize in a lane: use GPS, place car in lane, compensate for errors of GPS
   (GPS accuracy: several meters, accuracy needed: several cm)
5. **Path Planning**
   Run algorithms to perform path/motion planning, taking into consideration car kinematics, decide whether to stay in lane or switch lanes
6. **Navigation**
   After intensive computation, if decision is to take action, actuation in near-real time of vehicle controls to ensure safety and comfort

 21

Autonomous Driving Based on Deep Learning

Vehicle LiDAR Raw Data

Nvidia data collected with Quanergy LiDAR sensors

 22

QUANERGY Autonomous Driving Based on Deep Learning

Occupancy Map, Object Detection & Tracking

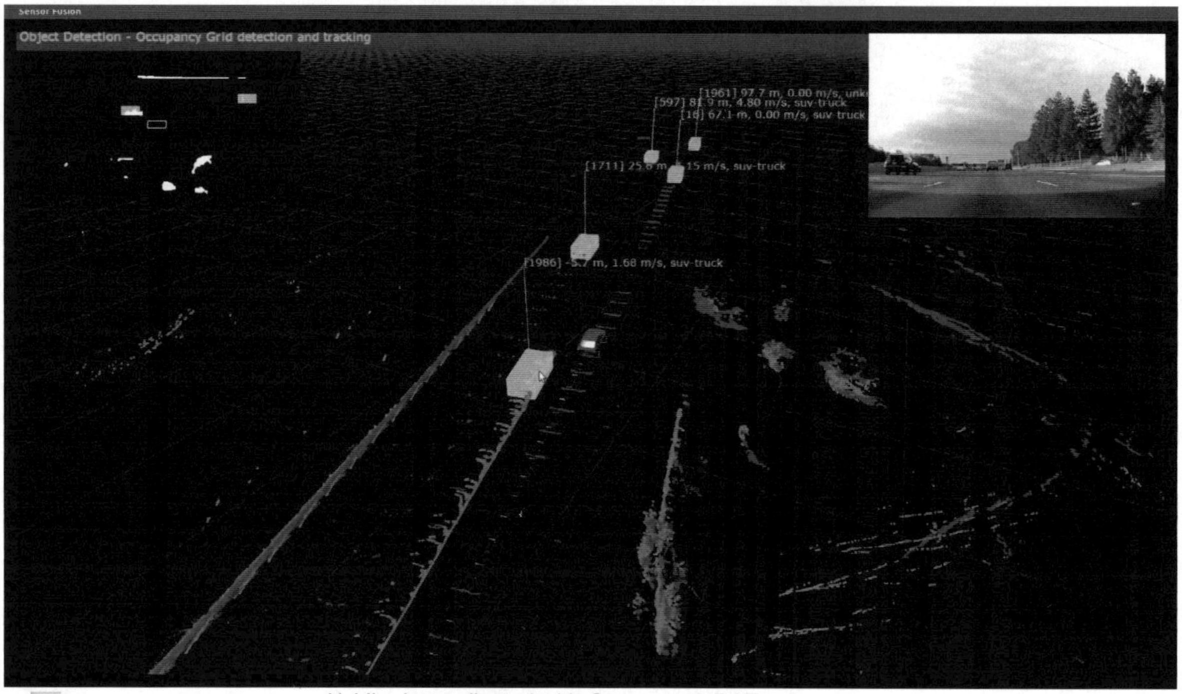

Nvidia data collected with Quanergy LiDAR sensors

23

QUANERGY Autonomous Driving Based on Deep Learning

Localization, Path Planning & Navigation

Nvidia data collected with Quanergy LiDAR sensors

24

QUANERGY Automotive Sensing and Perception

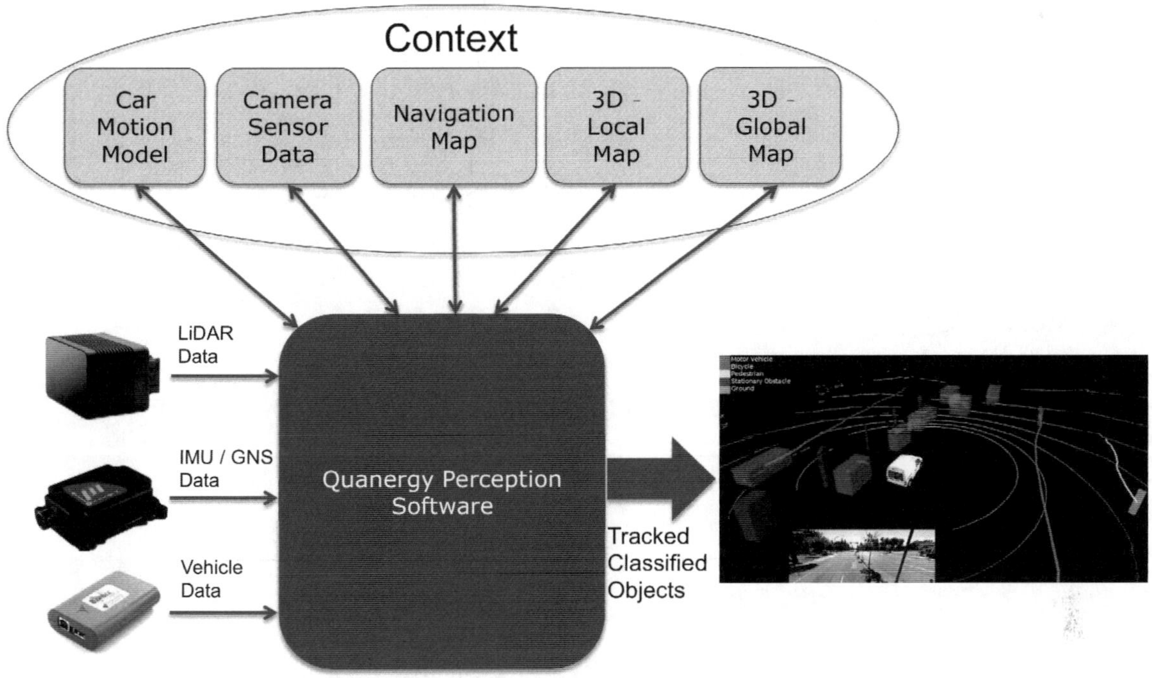

QUANERGY Levels of Automation for Vehicles

NHTSA Level	SAE Level	SAE Name	SAE Narrative Definition	Execution of Steering and Acceleration/ Deceleration	Monitoring of Driving Environment	Backup Performance of Dynamic Driving Task	System Capability (Driving Modes)
	Human driver monitors the driving environment						
0	0	**Non-Automated**	The full-time performance by the human driver of all aspects of the dynamic driving task, even when enhanced by warning or intervention systems	Human driver	Human driver	Human driver	n/a
1	1	**Assisted**	The driving mode-specific execution by a driver assistance system of either steering or acceleration/deceleration using information about the driving environment and with the expectation that the human driver perform all remaining aspects of the dynamic driving task	Human driver and system	Human driver	Human driver	Some Driving modes
2	2	**Partial Automation**	The driving mode-specific execution by one or more driver assistance systems of both steering and acceleration/ deceleration using information about the driving environment and with the expectation that the human driver perform all remaining aspects of the dynamic driving task	System	Human driver	Human driver	Some Driving modes
	Automated driving system ("system") monitors the driving environment						
3	3	**Conditional Automation**	The driving mode-specific performance by an automated driving system of all aspects of the dynamic driving task with the expectation that the human driver will respond appropriately to a request to intervene	System	System	Human driver	Some Driving modes
4	4	**High Automation**	The driving mode-specific performance by an automated driving system of all aspects of the dynamic driving task, even if a human driver does not respond appropriately to a request to intervene	System	System	System	Some Driving modes
	5	**Full Automation**	The full-time performance by an automated driving system of all aspects of the dynamic driving task under all roadway and environmental conditions that can be managed by a human driver	System	System	System	All driving modes

Source: Summary of SAE International's Draft Levels of Automation for On-Road Vehicles (July 2013)
NHTSA - National Highway Traffic Safety Administration
SAE - Society of Automotive Engineers

QUANERGY

Automation Deployment

- L1 platooning launches this year – fuel efficiency
- Automation up to L2 helps in all vehicles – accident reduction
- Skipping L3 for all vehicles could be wise – convenience in traffic jam assist; re-engaging the driver risky; does not improve productivity
- Passenger vehicles owned by consumers: jump from L2 to L4 (productivity), then L5 (cost savings); if L3 is used, auto-pilot naming should be avoided (Audi, BMW, Daimler, China)
- Trucks and Ride-sharing Fleets: skip from L2 to L5 – cost savings not realized until vehicles are driverless
- Starting L4/L5 applications are confined work sites
- Mining/Agriculture L5 vehicles used today (CAT)
- L5 for on-road 10+ years away; self-driving lanes could accelerate
- Benefits:
  - L1 – cost reduction (fuel)
  - L2 – safety
  - L3 – safety, convenience (not to be confused with comfort or productivity)
  - L4 – safety, comfort, productivity
  - L5 – safety, cost reduction (eliminate driver)
- Responsible and robust implemented of all automation levels have LiDAR as primary sensor

QUANERGY

Some Public Partners

QUANERGY

Sensata Technologies | QUANERGY

Sensata and Quanergy Strategic Partnership

In March 2016, Sensata and Quanergy announced strategic partnership:

- Exclusive partnership for the ground transportation market
- Companies to jointly develop, manufacture, and sell component level solid-state LiDAR sensors
- Quanergy to lead technology development including perception software
- Sensata to design for manufacture, make, and sell to ground transportation customers
- Sensata investment in Quanergy and on board of directors

Sensata Technologies

QUANERGY

QUANERGY

Thank You

Quanergy Sites:
Sunnyvale (HQ)
Detroit
Ottawa
Washington, D.C.
London
Munich
Dubai
Shanghai
Tokyo

Louay Eldada
+1-408-245-9500
louay.eldada@quanergy.com
www.quanergy.com

Princeton Lightwave

GeigerCruizer™
The Eye-Safe, Scalable, 250 meter Automotive LIDAR

Sam Wilton

1

The Geiger Mode Advantage

Princeton Lightwave

- long range (> 300 meters)
- high resolution (< 0.05 degrees)
- fast frame rate (> 24 Hz)
- enhanced eye-safety (0.1 W / cm^2)
- additional ambient intensity (> 100,000x dynamic range)
- reliable performance in all weather (sun, rain, snow, fog)
- extreme performance in ideal weather (cloudy, night)
- continuous digital data acquisition enables deep level DSP

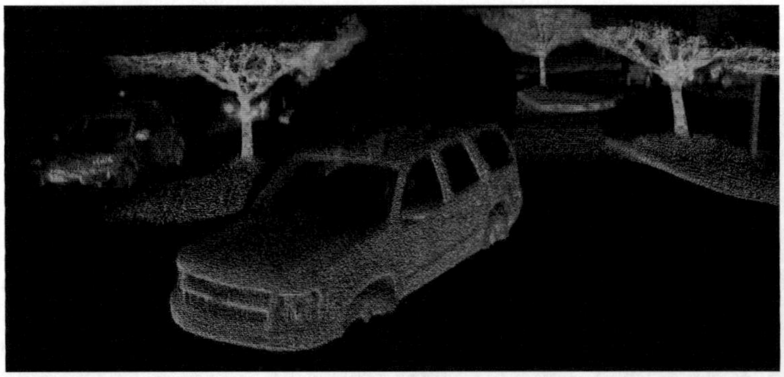

2

GeigerCruizer™ Performance Specifications

GeigerCruizer LR
- 200 meter range (10% reflectance)
- 512 x 128 pixels (2 mrad)
- 60 x 15 degree fov
- 24 Hz image rate

GeigerCruizer Demonstrator Prototype
- 200 meter range (10% reflectance)
- 512 x 64 pixels (2 mrad)
- 50 x 7.5 degree fov
- 12 Hz image rate

Long Range Detection

Geiger mode detectors are triggered by single photons, allowing low reflectance targets to be detected at several hundred meters range with only nanojoules of laser pulse energy.

e.g. 10% reflectance

	100 m	200 m	300 m	400 m	500 m	600 m
Tx 10 nJ						
Rx 10 mm	*15 photons*	*4 photons*	*2 photons*	*1 photons*	*0.6 photons*	*0.4 photons*
Probability ≥ 1 photon	*99.999%*	*98%*	*82%*	*63%*	*45%*	*33%*

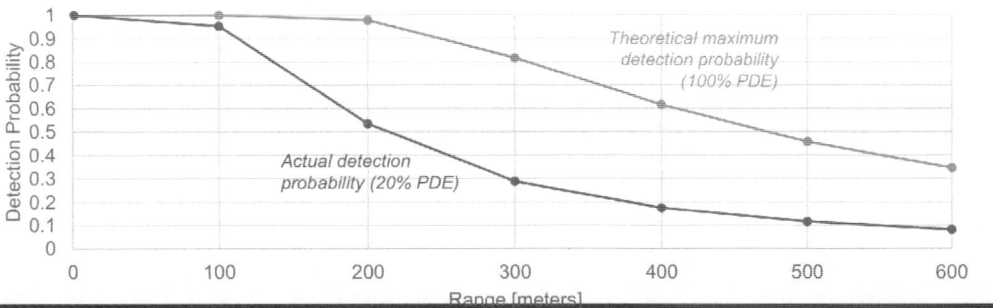

GeigerCruizer™ Demo > 300 meter range

- The combination of Geiger mode single photon sensitivity and eye-safe SWIR wavelengths enables >300 meter operating range, making the GeigerCruizer LIDAR ideal for highway driving conditions.
- Vehicles and road-signs contain retroreflective elements, allowing them to be seen at distances over 600 meters.

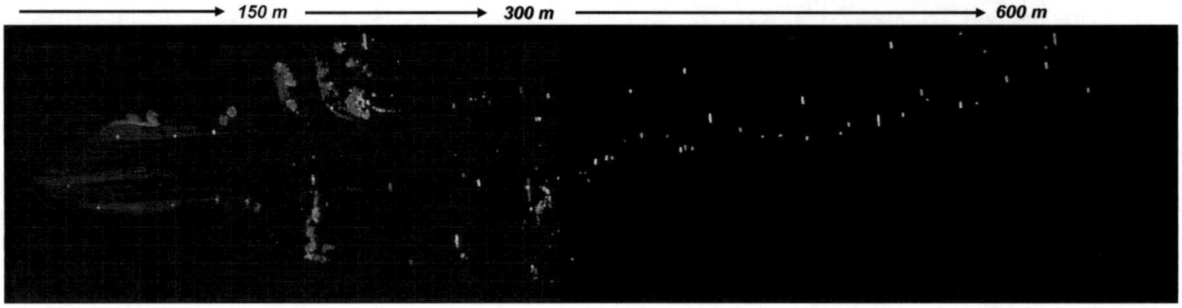

Scalable Resolution With No Gaps

Geiger mode LIDAR systems generate a continuum of raw time-of-flight data points from a series of discrete laser pulses received into a focal plane array of GmAPD detectors.

Raw data is spatiotemporally integrated to generate resolution elements of any desired shape and size from the same source information. Software foveation is used to dynamically adjust angular resolution based on environmental conditions and scene roi

1 point integration
- 1 bit dynamic range
- random, noisy returns
- highest spatial resolution
- no confidence level

64 point integration
- 6 bit dynamic range
- lower precision returns
- higher spatial resolution
- lesser statistical confidence

256 point integration
- 8 bit dynamic range
- higher precision returns
- lower spatial resolution
- greater statistical confidence

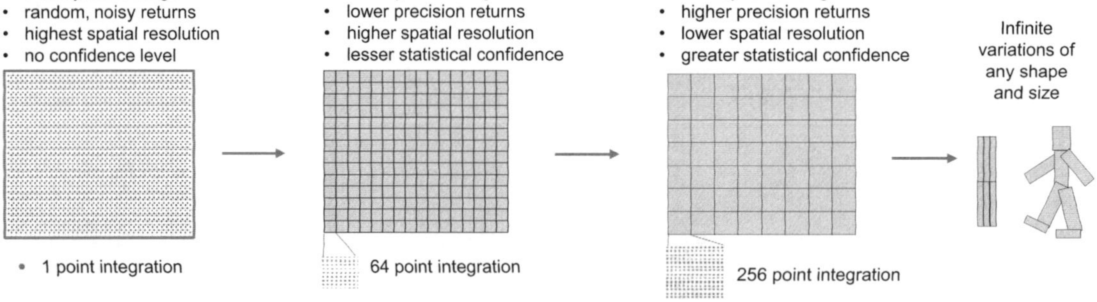

Windowing processes are used to process overlapping sections of raw data to increase perceived resolution with no additional cost of laser and/or detector resources. The maximum number of window centers is equal to the number of raw data points. GeigerCruizer™ acquires 20480 x 512 raw data points at 24 Hz. With adequate DSP resources, it is possible to generate 10 Megapixel range images.

GeigerCruizer™ Demo Resolution Scaling

8 mrad

Similar to Velodyne HDL-64

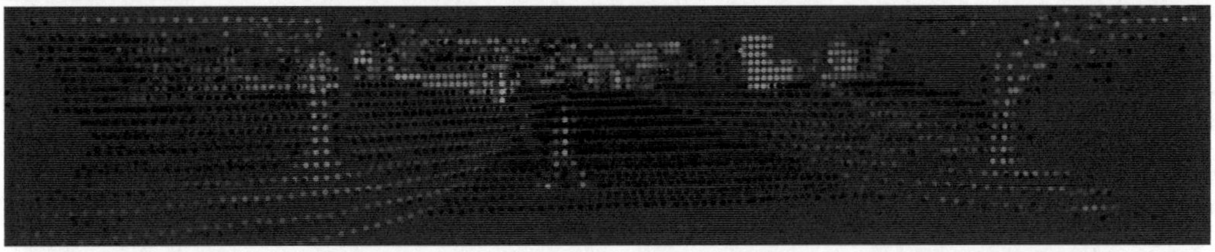

4 mrad

1 mrad

7

High Frame Rate

Large format FPAs allow thousands of raw data points to be acquired per microsecond. Discrete laser pulses are temporally windowed to enhance perceived image rate. Maximum image rate is bounded by laser pulse frequency.

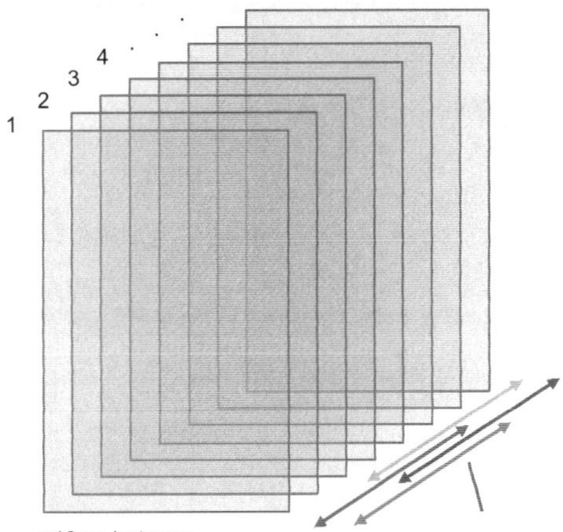

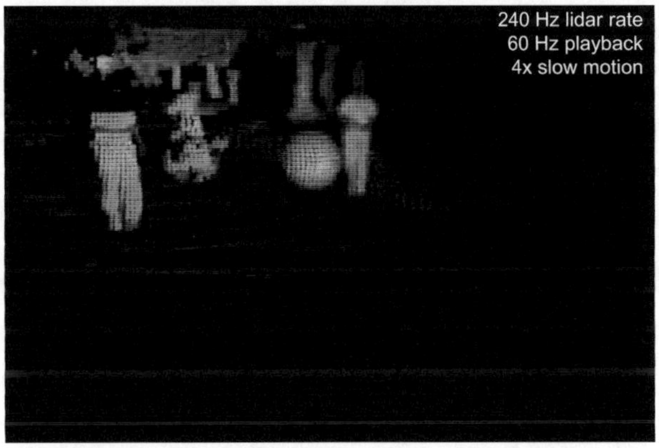

pulse group windowing techniques can be used for temporal resolution enhancement.

8

Signal and Ambient Intensity

Geiger mode detectors cannot directly disambiguate between individual photons from a laser pulse and ambient light of the same wavelength.

Although ambient light produces noise in the TOF data, we take advantage of this additional information to generate ambient intensity images. In this way, the sensor can act as both a LIDAR and a single-photon sensitive passive SWIR camera, simultaneously.

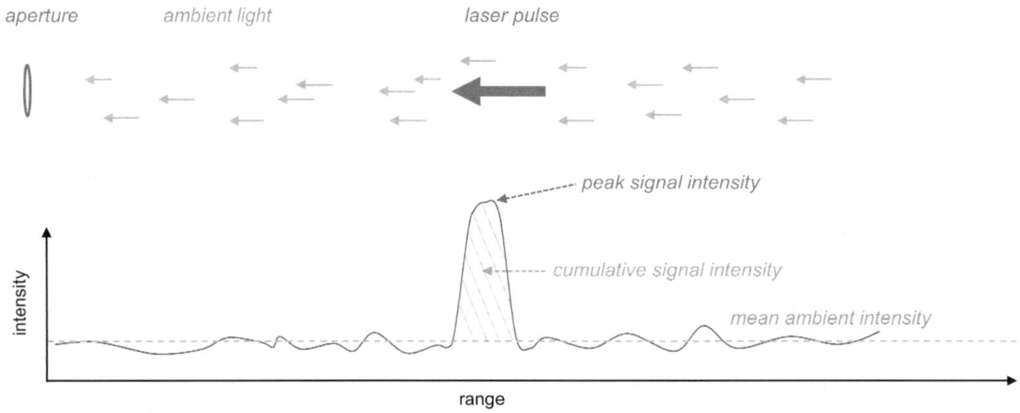

9

GeigerCruizer™ Demo Intensity

- Ambient intensity images provide additional information for intelligent video analytics of Geiger mode LIDAR data.
- Depending on environmental conditions, it is possible for the dynamic range of ambient intensity images to exceed the dynamic range of signal intensity images due to statistical dependence on both quantity and time-delay between avalanches, rather than quantity alone.
- Maximum dynamic range is dependent on data density, exposure time, dark noise rate, and time-bin width. Theoretically, it is possible to obtain ambient intensity images with a dynamic range greater than 100,000x (~100 kHz to 10 GHz photon flux rate).

10

Reliable in Inclement Weather

The majority of weather obscurants are a result of water or ice particulates in the atmosphere, which cause both backscattering and forward-scattering of light.

- backscattering in the transmit path causes signal attenuation and non-linear noise
- backscattering in the return path causes signal attenuation

- forward-scattering in the transmit path causes signal attenuation
- forward-scattering in the return path causes signal attenuation and blooming

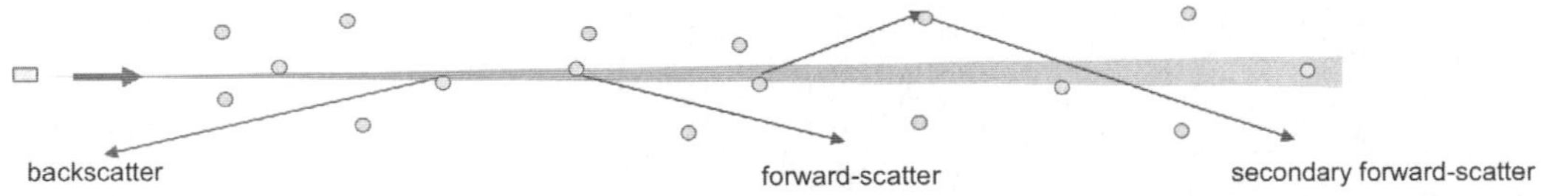

- Single photon sensitivity allows the GeigerCruizer LIDAR to see low reflectivity targets at distances greater than 100 meters, even in inclement weather conditions.
- Advanced, dynamic, line-of-sight noise-filtering algorithms are able to filter backscatter without affecting point cloud quality.

11

GeigerCruizer™ Demo in Rain

12

GeigerCruizer™ Demo in Snow

Snow gusts are particularly challenging due to the small grain size of the ice crystals, which cause strong scattering of signal photons.

The range to the highway is 200 meters, keep an eye out for the truck

GeigerCruizer™ Demo in Fog

- Fog is the most challenging atmospheric obscurant due to its very small particle size and total volumetric permeation.
- Imaging through fog will always degrade the quality of LIDAR data, but in this real world example, we demonstrate > 100 meter range.
- Beyond 100 meters, objects are a bit fuzzy, but sometimes cars can be seen at 200 meters

100 meters

*videos not synchronized

GeigerCruizer™ Demo with Window Obscuration

Princeton Lightwave

Our system architecture is highly resistant to front window obscuration. Virtually zero performance degradation is observable between a clean, dirty, and water sprayed windows.

15

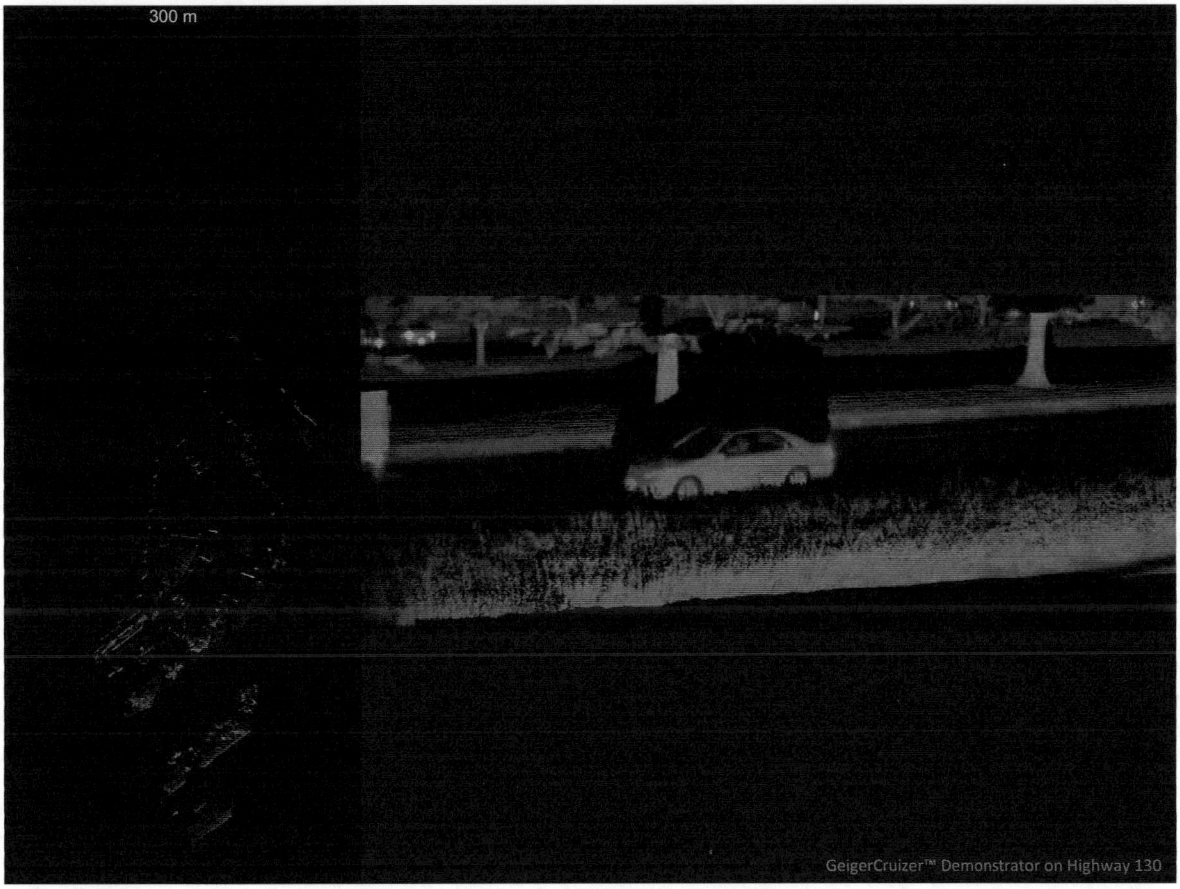

NORTHERN CALIFORNIA, NEVADA, AND UTAH (NCNU)

SUNA TAYMAZ, VICE-PRESIDENT, CORPORATE STRATEGY

1914

INSURANCE BUREAU

TRAVEL BUREAU

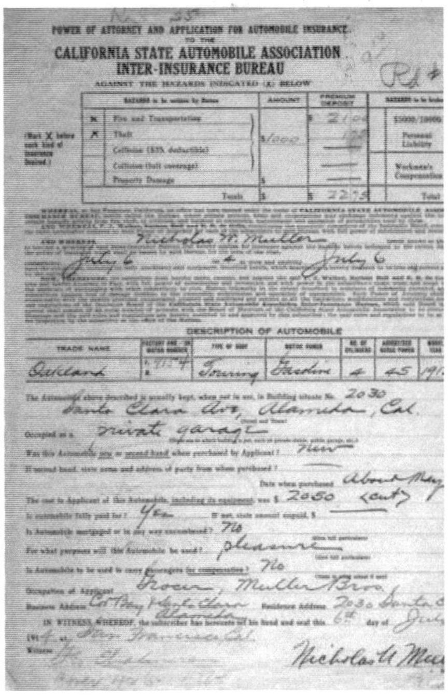

POWER OF ATTORNEY AND APPLICATION FOR AUTOMOBILE INSURANCE
TO THE
CALIFORNIA STATE AUTOMOBILE ASSOCIATION
INTER-INSURANCE BUREAU
AGAINST THE HAZARDS INDICATED (X) BELOW

DESCRIPTION OF AUTOMOBILE

ROAD SIGNAGE

DMV
REGISTRATION

AAA ADOPTED WHICH INNOVATION(S) EARLY?

CAR-SHARING

SAFETY TECHNOLOGY SIMULATIONS

LIVE ROAD TRAFFIC CONDITIONS

MOBILE ENGINE DIAGNOSTICS

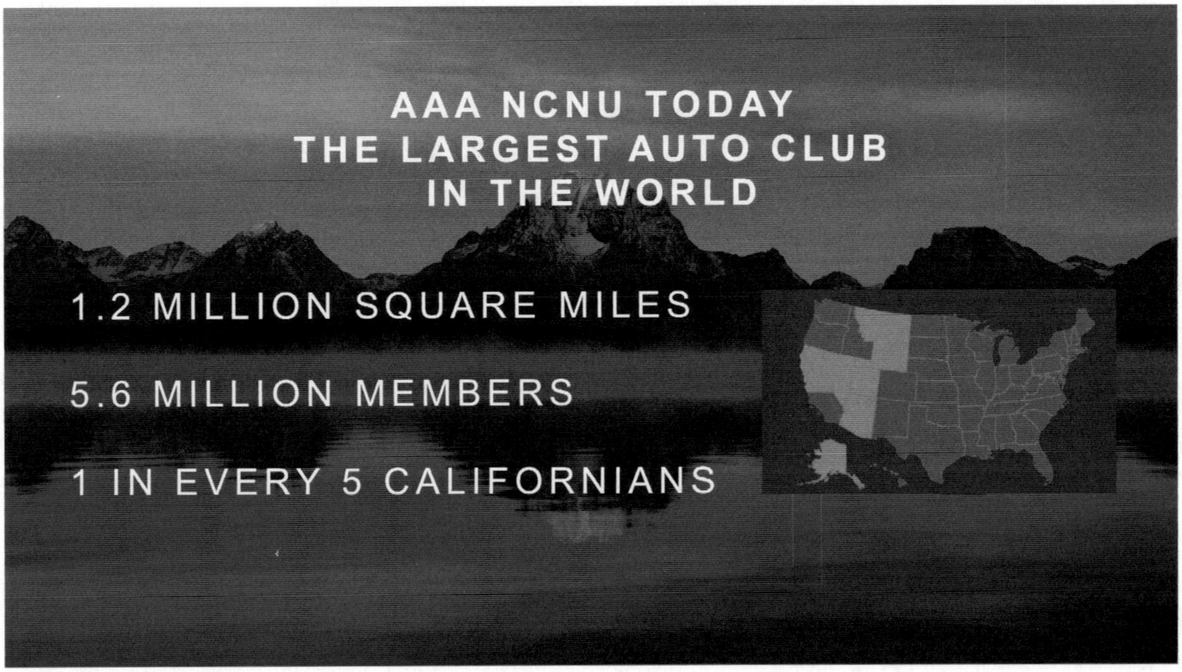

AAA IN THE US AND ABROAD

~30 FEDERATED US CLUBS

60+ INTERNATIONAL CLUBS

OUR WORLD IS CHANGING...

DISRUPTIVE TRENDS

CAR OWNERSHIP

30% of Millennials do not plan to purchase a car in the near future

CAR USAGE

The average car is in motion for only 60 minutes per day

CAR SHARING | **RIDE SHARING**

The car sharing and ride sharing market is predicted to exceed $30B by 2024

MILLENNIAL TRENDS

23% 14% 65%

DECLINING EMERGENCY ROAD SERVICE USAGE

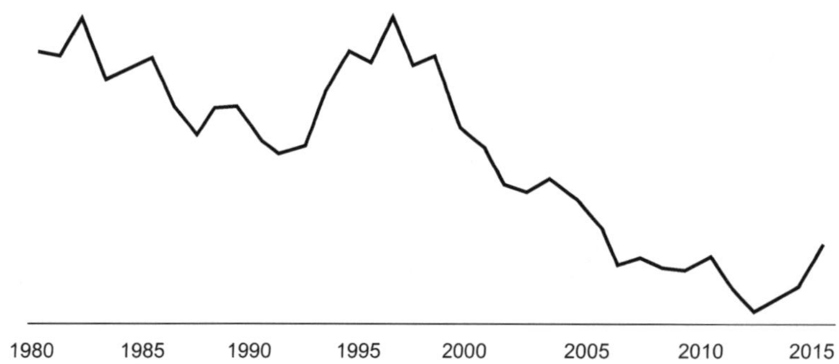

Source: AAA National ERS usage statistics

OUR VISION FOR TRANSPORTATION
IS TO ACCELERATE THE AUTONOMOUS, CONNECTED, ELECTRIC AND SHARED FUTURE

MID-1950's ADVERTISING
MERCEDES-BENZ CONCEPT

AAA SURVEY ON AUTONOMOUS VEHICLES

- **75%** of U.S. drivers would be afraid to ride in a self-driving vehicle
- **54%** would feel less safe sharing the road with self-driving cars
- **59%** want autonomous technology in the next vehicle they buy or lease, while 25% would not want the technology

THE CAR ECONOMY

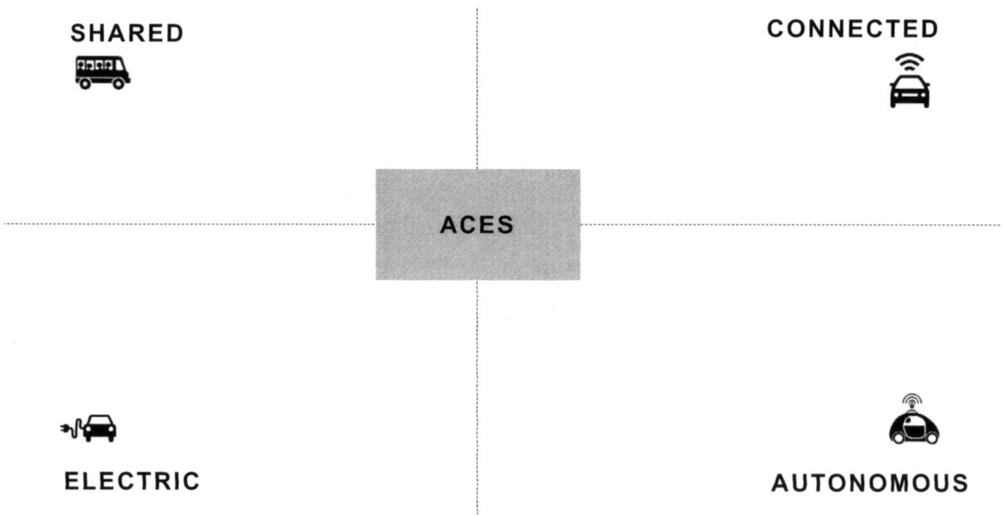

A KEY POINT OF ENTRY

MOBILE REFUELING: NEW OPPORTUNITIES

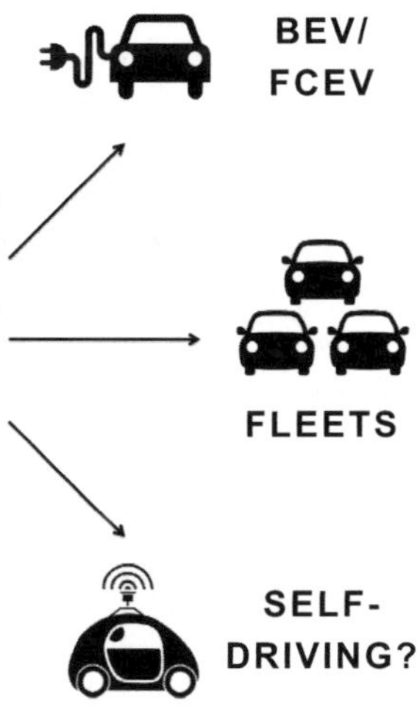

CONNECTED CARS

SMARTtrek (ARIZONA ONLY)

- SIMPLE ONLINE DASHBOARD
- REAL-TIME VEHICLE HEALTH ALERTS
- DRIVING FEEDBACK
- MAINTENANCE REMINDERS

AUTONOMOUS

FOCUS ON OUR MEMBERS

ACCELERATE THE TRANSITION

UNCOMPROMISING SAFETY

OUR ROLE IN THE FUTURE OF MOBILITY

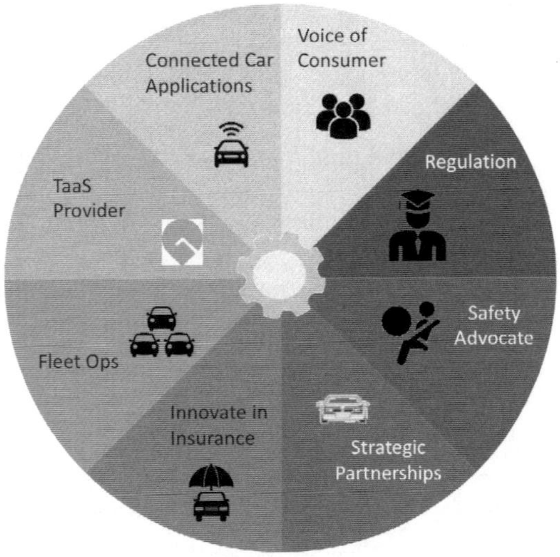

Simulationstechnologie GmbH
MSC Software Company

VTD

VIRES Virtual Test Drive®

Marius Dupuis
VIRES Simulationstechnologie GmbH

VIRES at a Glance - **Overview**

Focus on key product **VIRES Virtual Test Drive®**

Software tool-chain supporting the development, test and validation of automated driving, driver assistance and active safety systems

Components

- traffic simulation
- scenario simulation
- image generation
- sensor simulation / stimulation
- database design tools
- SDKs
- etc.

VTD

VTD is Linux-only

Partner in **Standardization Initiatives**

Open*DRIVE*
Open*CRG*
Open*SCENARIO*

VIRES – Products and Services

VIRES Simulationstechnologie GmbH, MSC Software Company

Products

VTD VIRES Virtual Test Drive®

Purpose	Module	Extension	Industries
Key Task	Task Control (Framework)		
Scenario	Traffic Simulation Scenario Editor		
Visualization and Sensors	Image Generator v-IG	SDK for Render Plug-ins and ray-traced sensors	
Sensors, Dynamics Interfacing	Module Manager	SDK for ObjSensors, Dynamics, Custom Mod.	
high freq. Road Contact	OdrGateway		
GUI	Operator Station		
sound simulation	Sound		
database editor	Road Designer		
content	ready-to-go databases		
road eval. library	Odr Manager		

Services

Name	Purpose	Industries
on-site integration	make VTD fit for an application	
adaptation	custom and core extensions	
user training	make the user fit for VTD	
custom-built databases and scenarios	content	

OpenDRIVE, OpenCRG, OpenSCENARIO:

- Participation in Specification Efforts
- User Support
- Tool Development

Projects

Name	Purpose	Industries
PEGASUS	German research project "automated driving"	
Enable-S3	European research project "automation"	

VIRES Simulationstechnologie GmbH, MSC Software Company

Motivation

Motivation – *Reality*

VIRES

Knowledge and Perception

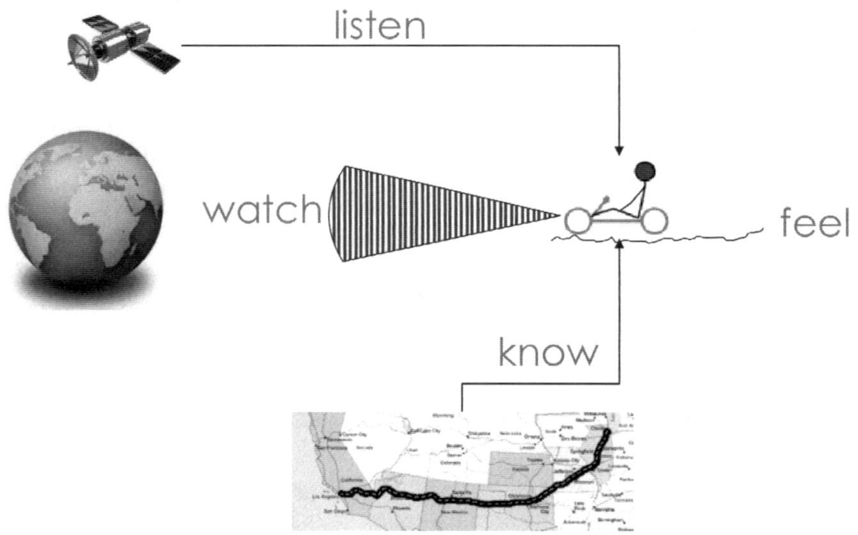

Motivation – *Preparing for Reality*

VIRES

Ground Truth and Models

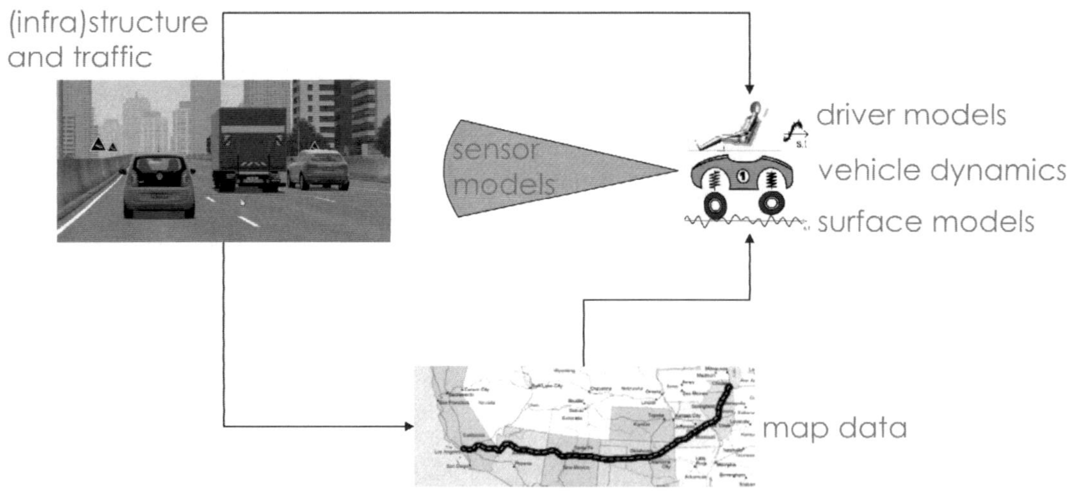

Motivation – The Challenge

Similarity!

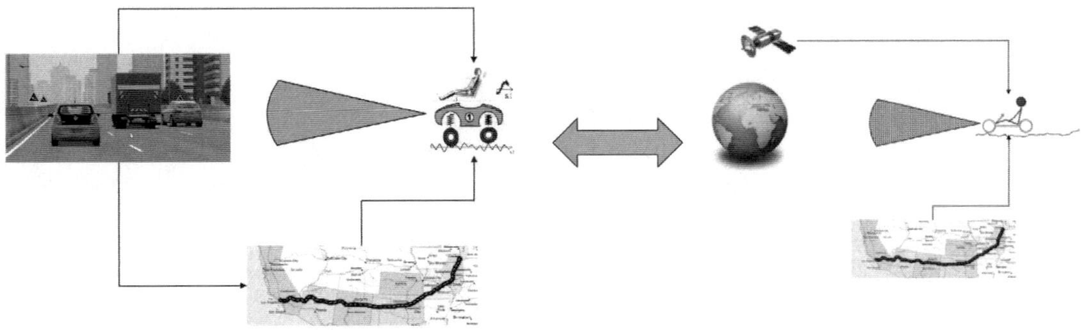

VTD 07/12/2017 (c) 2017 VIRES GmbH, protection notice DIN ISO 16016 Slide 7

Our Role

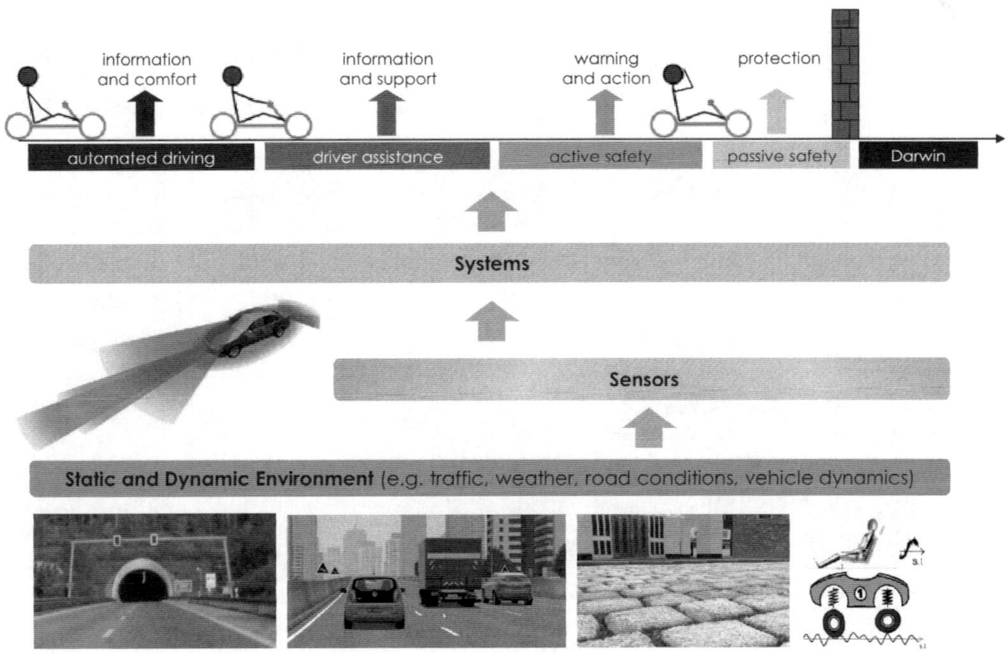

VTD 07/12/2017 (c) 2017 VIRES GmbH, protection notice DIN ISO 16016 Slide 8

Objectives

Objectives – *Coherent Development*

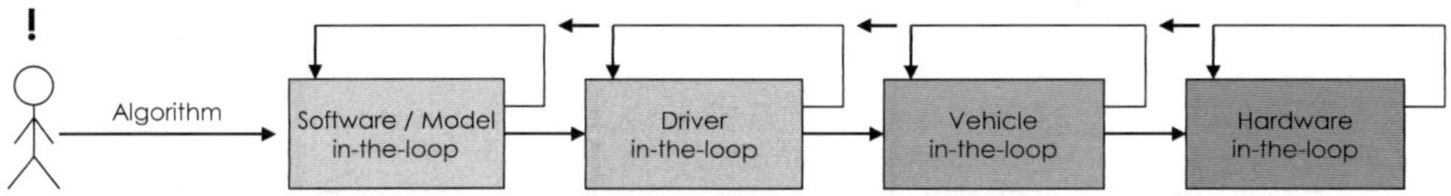

setup				
interaction	autonomous	interactive	interactive	autonomous
number of tests	large	moderate	small	medium
speed	non real-time	real-time	real-time	real-time
Driver	virtual	real	real	virtual
Dynamics	virtual	virtual	real	virtual
Sensors	virtual	virtual	virtual / real	virtual / real
Systems	virtual	virtual	virtual / real	real
Traffic	virtual	virtual	virtual / real	virtual
portability and compatibility	Modules	Configuration	Record Data	Scenarios

Objectives – *Structured Development*

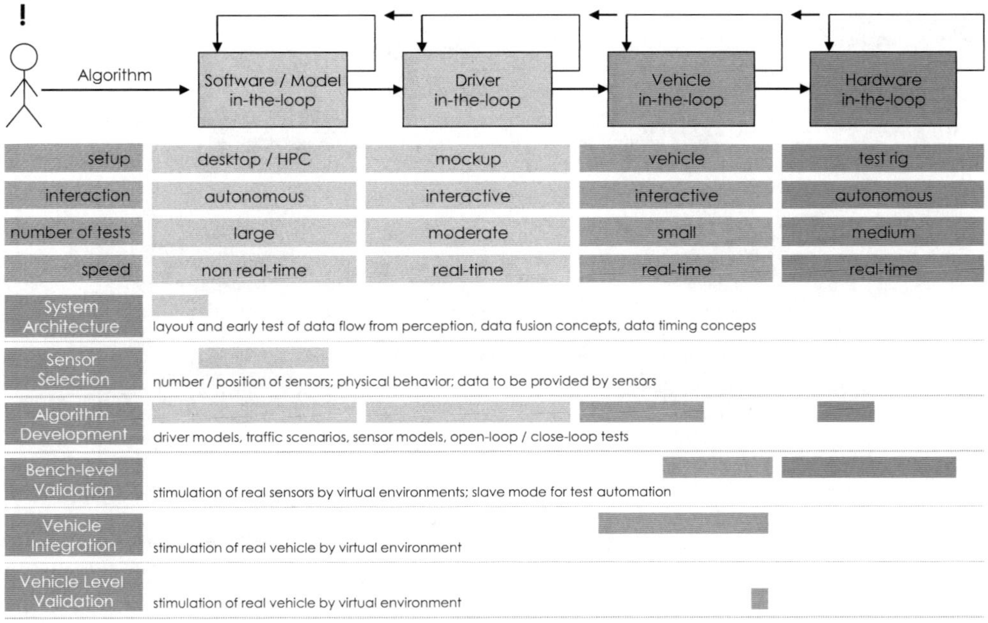

Workflow

Workflow – Overview

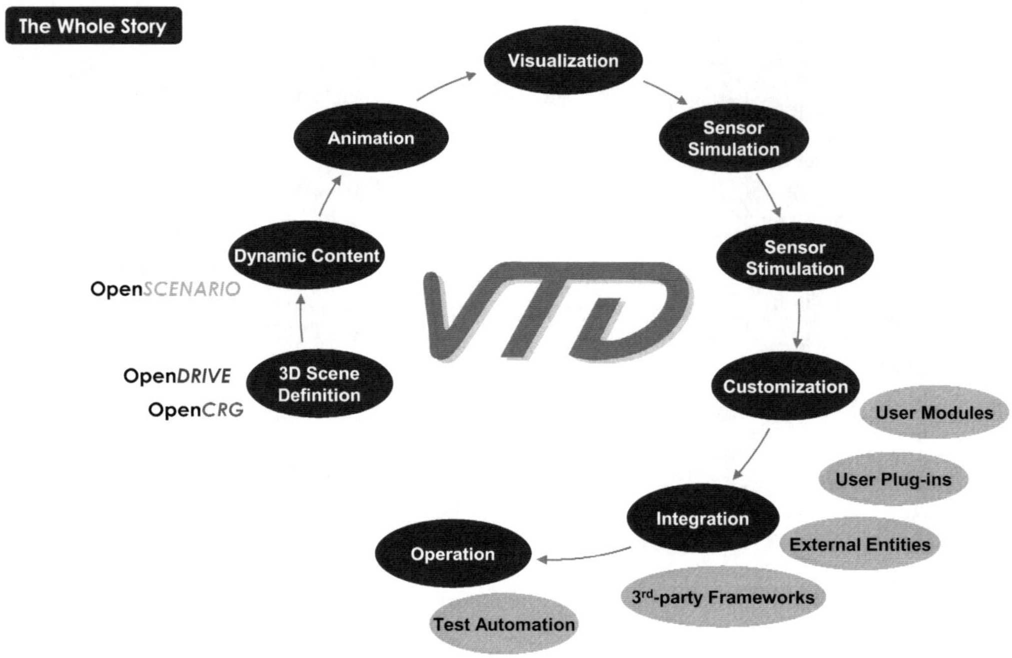

07/12/2017 (c) 2017 VIRES GmbH, protection notice DIN ISO 16016 Slide 13

Workflow – Overview

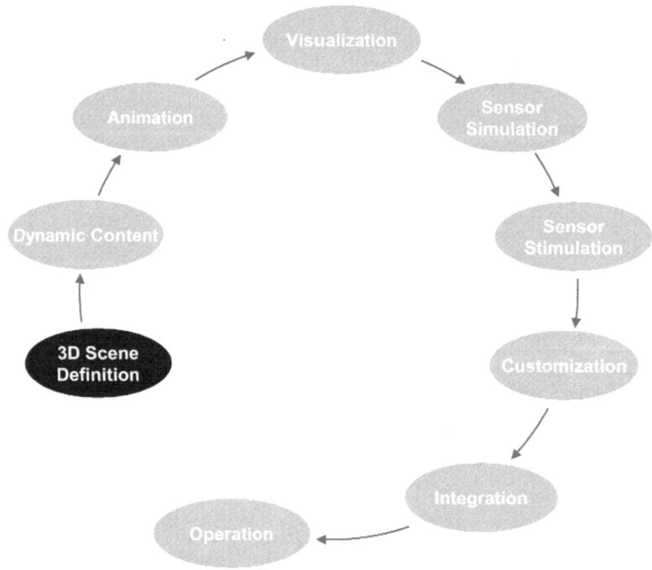

07/12/2017 (c) 2017 VIRES GmbH, protection notice DIN ISO 16016 Slide 14

Workflow – *Environment Simulation*

3D-World

VTD 07/12/2017 Slide 15

Workflow – *Environment Simulation*

3D-World – Resembling Reality

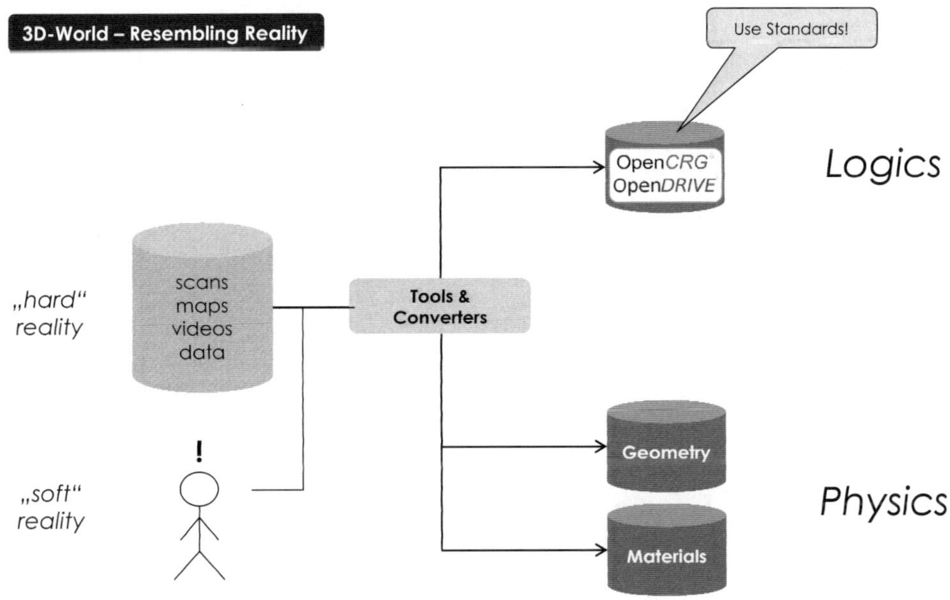

VTD 07/12/2017 Slide 16

Workflow – *Environment Simulation*

VIRES

3D World

- **generic** vs. **geo-specific**
- **country-coded** signs / signals / layouts
- analytical **ground-truth** data in **OpenDRIVE**
- **GPS** reference for localization
- **road surface** details

Workflow – *Overview*

VIRES

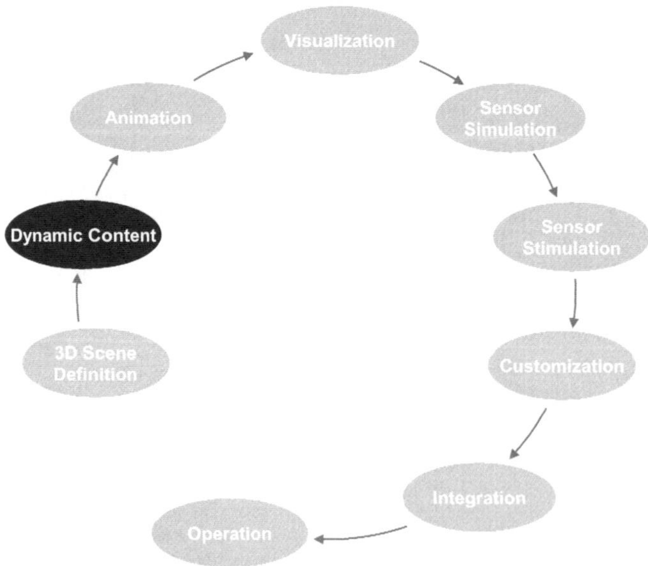

Workflow – *Environment Simulation*

Scenarios – Sampling Reality

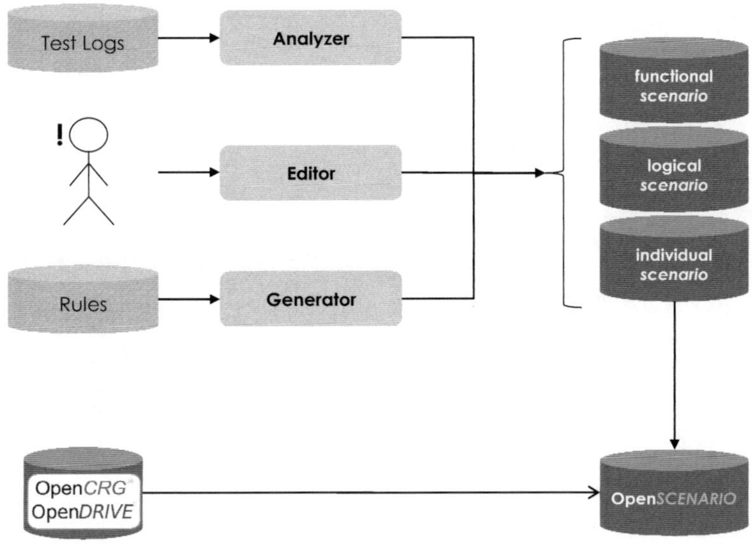

Workflow – *Overview*

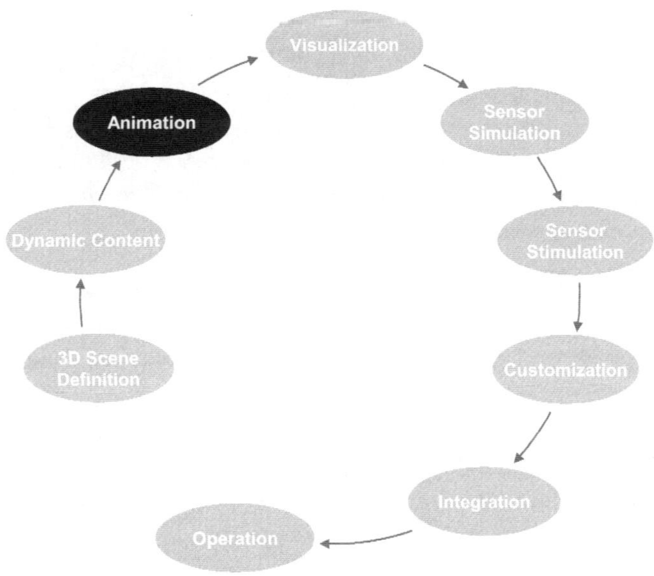

Workflow – *Environment Simulation*

Animation – Resembling Situations

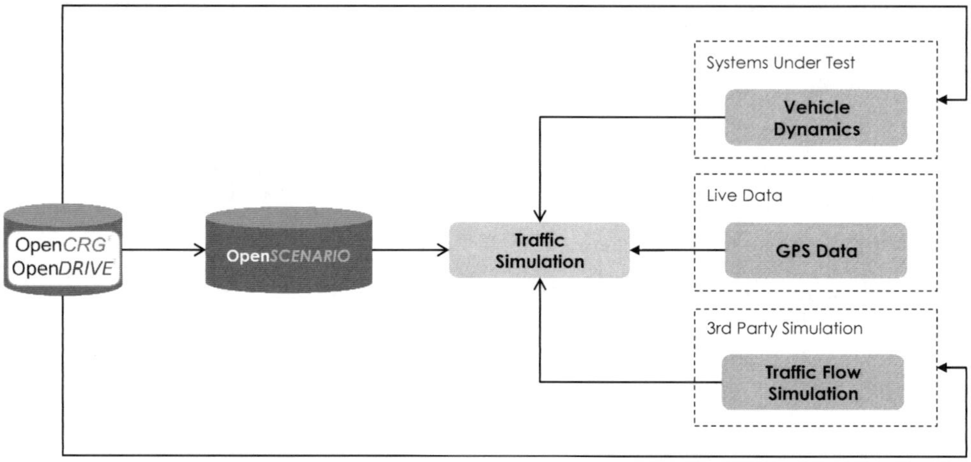

Workflow – *Traffic Simulation*

Traffic

- **autonomous** vs. **deterministic** behavior
- realistic traffic **density** and **features**
- **microscopic** and **nanoscopic** context
- unlimited number **of external entities**
- configurable **driver models**
- configurable **vehicle dynamics**
- **pedestrian** simulation

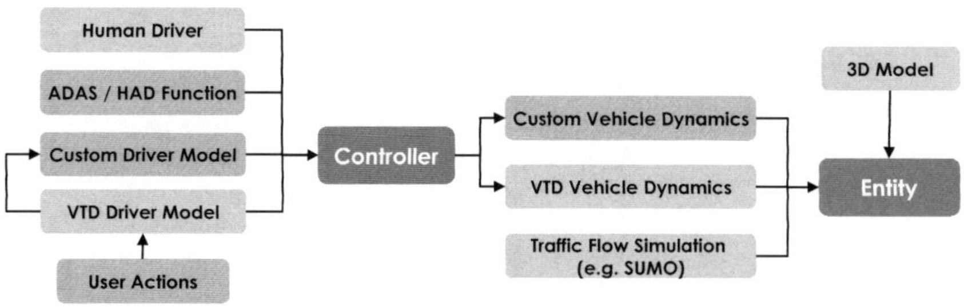

Workflow – Overview

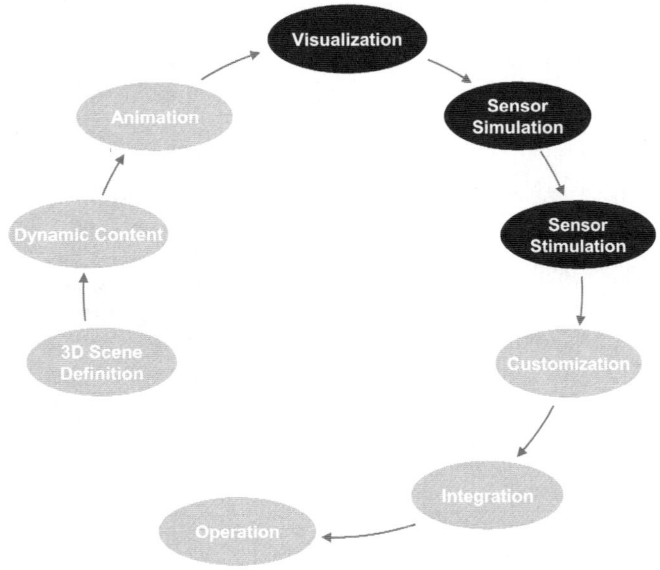

Workflow – Perception

Animation – Resembling Situations

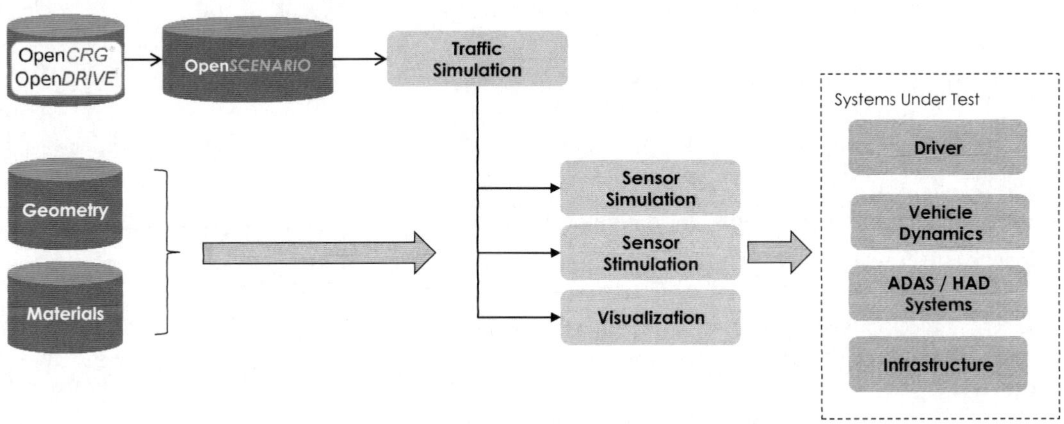

Workflow – Perception

Visualization – Seeing Content

- **low / high dynamic range**
- **real-time** performance
- image **output** on monitor / network / shared memory
- customizable via **SDK**
- customizable **material tables**
- **single-/multi-channel** installation
- configurable **distortion** and **blending**
- **fish-eye** capabilities
- various **weather** and **light** conditions

Workflow – Perception

Visualization – For Engineers

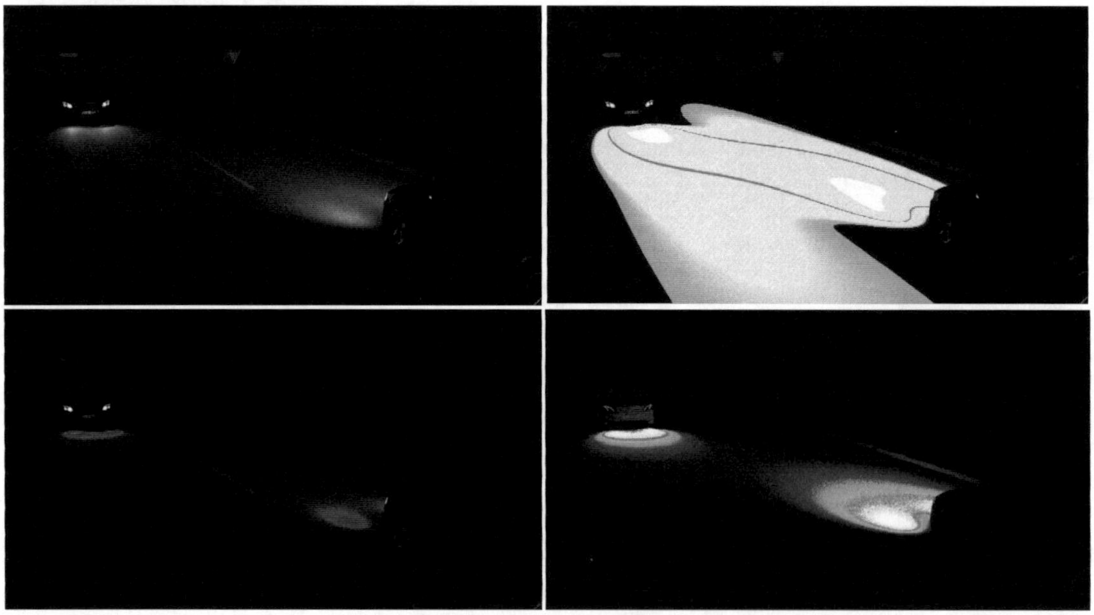

Workflow – Perception

Visualization – For Eyes and Cameras

Workflow – Perception

Visualization – For Eyes and Cameras

Workflow – *Perception*

Sensor S(t)imulation - Concepts

- **object lists** vs. **raw data** (logics vs. physics)
- **ground-truth** data
- **real-time** performance (and faster)
- multiple parallel instances
- **headless** simulation (e.g. for HPC)

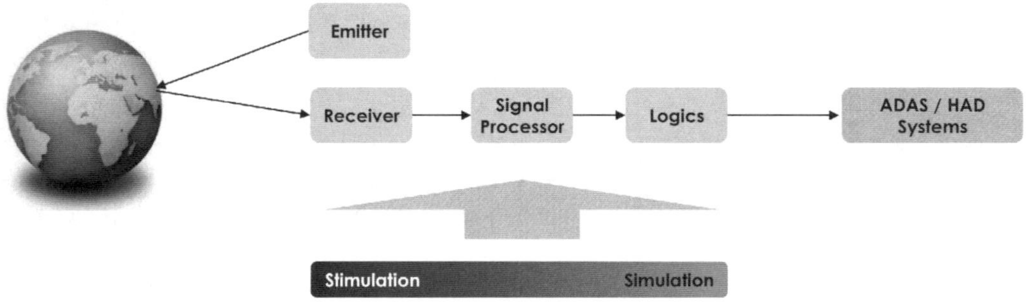

Workflow – *Perception*

Sensor Simulation - Concepts

Workflow – Perception

Sensor Simulation – Providing Consistency

VTD 07/12/2017 Slide 31

Workflow – Perception

Sensor Simulation – Implicit Ground Truth

VTD 07/12/2017 Slide 32

Architecture

Architecture – *VIRES Virtual Test Drive*

VIRES

Rigid Core – Soft Shells

- **one** simulation **core** for all stages and environments
  - inherent **compatibility** of key simulation data
- **customization** on various levels
- **integration** into existing setups
- **open interfaces** for interaction with 3rd party tools
- **real-time** and non-real-time operation
- **master / slave** modes (for co-simulation)
- use of (de-facto) **standards**
  - OpenDRIVE (road logics)
  - OpenCRG (road surface)
  - OpenSCENARIO (within 2017)
  - OpenSceneGraph and OptiX (graphics)
  - IES (headlight distribution)

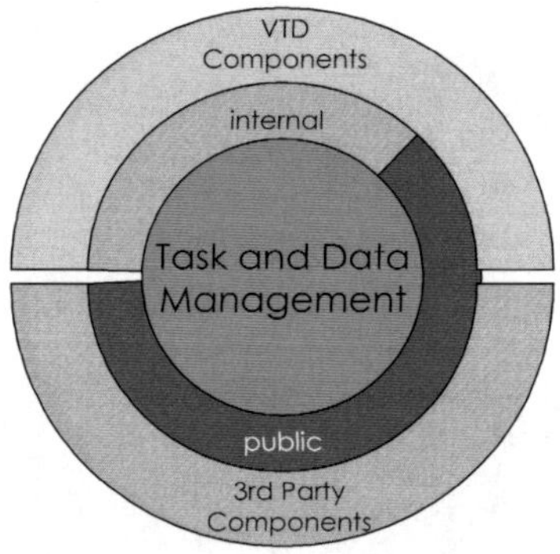

(c) 2017 VIRES GmbH, protection notice DIN ISO 16016

Thank you!

Drive Sweden
- an update on Swedish Automation Activities
Jan Hellaker
Program Director
www.drivesweden.net/en
Twitter: @DriveSweden
DRIVE SWEDEN

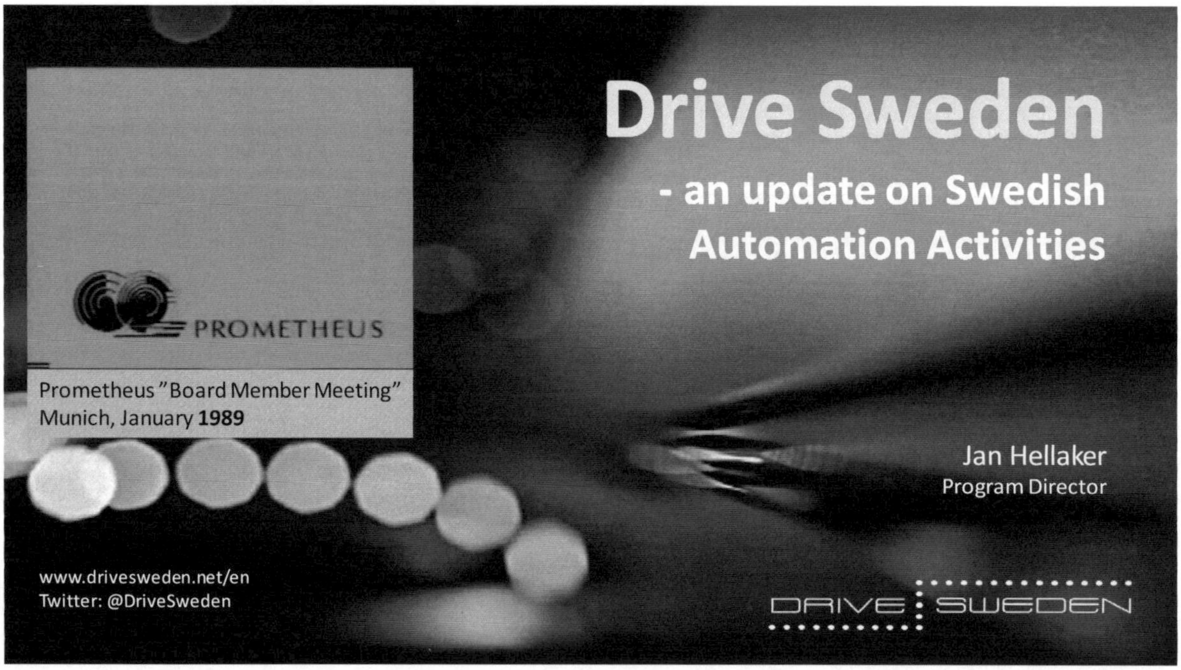
PROMETHEUS
Prometheus "Board Member Meeting"
Munich, January 1989
Drive Sweden
- an update on Swedish Automation Activities
Jan Hellaker
Program Director
www.drivesweden.net/en
Twitter: @DriveSweden
DRIVE SWEDEN

SWEDEN

VOLVO

SCANIA

VOLVO

ZENUITY

Make it real.

LYNK&CO

EINRIDE

DRIVE SWEDEN

Drive Sweden

Strategic Innovation Program

A government-sponsored cross-functional collaboration platform aiming at designing the next generation mobility system for people and goods; based on connected, automated and shared vehicles.

https://youtu.be/WmYsWYDQxul

VINNOVA

formas

Forskningsrådet Formas

Swedish Energy Agency

DRIVE SWEDEN

TRAFIKVERKET
TRANSPORT STYRELSEN

VOLVO
SCANIA
NEVS
nauya
LOCAL MOTORS
UNITI

Göteborgs Stad
Stockholms stad
kista SCIENCE CITY

KTH
vti
CHALMERS
LiU LINKÖPINGS UNIVERSITET
HÖGSKOLAN I HALMSTAD
RISE

ERICSSON
SAAB
IBM
Autoliv
RideCell
BESTMILE
MINDCONNECT
CONSAT
UNIQUE
veridict
carmenta
swarco
dynniq energising mobility
ApParkingspot

telenor connexion
TELE2
Telia

Norconsult
WSP
ÅF
ATKINS
SWECO
ch2m

ASTAZERO
ICEMAKERS

Sustainable innovation
Forum for transport innovation
GOOGOL

LINDHOLMEN SCIENCE PARK

Nobina
Samtrafiken
CABONLINE
Hertz
UBER
UbiGo
Freelway
AUTONOMOUS MOBILITY
transdev
postnord

DRIVE SWEDEN

Sweden4Platooning

FFI Fordonsstrategisk Forskning och Innovation

- Three year project (2017-19) with Scania, Volvo Group, Royal Institute of Technology, SICS Swedish ICT, DB Schenker AB and the Swedish Transport Administration
- Main deliverables:
  - Pilot Multi-brand C-ACC (longitudinal control) in a haulage company operation
  - Demonstrate Multi-brand platooning (lateral and longitudinal control) at a test track

DRIVE SWEDEN

- **Four-truck platoon**
  - Drayage application
  - 1 L0 + 3 L4 trucks
  - 80 km/h
  - 10-12 m gap

MINISTRY OF TRANSPORT
CONNECTING SINGAPORE

PSA
The World's Port of Call

SCANIA

Volvo Group

VOLVO

The 'T-pod' is an electric, driverless and/or remotely controlled vehicle, set for prototype testing in late 2017.

Cargo capacity: 15 standard pallets
Gross vehicle weight: 20 tons
Length: 7 meters (23 feet)
Battery capacity: 200 kWh
Range: 200km (124 miles)

EINRIDE

NEVS

NEVS' vision is to shape mobility on product, service and system levels

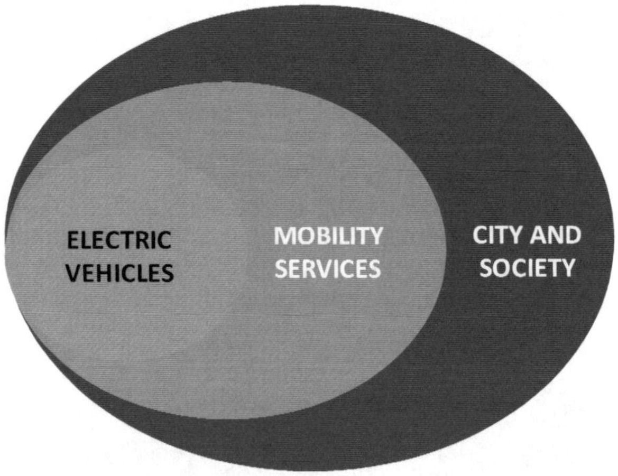

MOBILITY SERVICES

Car sharing and connectivity solutions
Integrated mobility services

NEVS

InMotion concept vehicle at CES Asia

MoU with Didi Chuxing

100 families to commute by self-driving Volvo XC90s. on public roads in Gothenburg, Sweden.

- Pilot will begin in Q4 of 2017
- Highly-automated driving (L4) on demand
- No safety driver
- Secondary tasks allowed
- Certified roads only, ~50 km
- Weather limitations

One of the test cars

Drive Me

SELF-DRIVING CARS FOR SUSTAINABLE MOBILITY

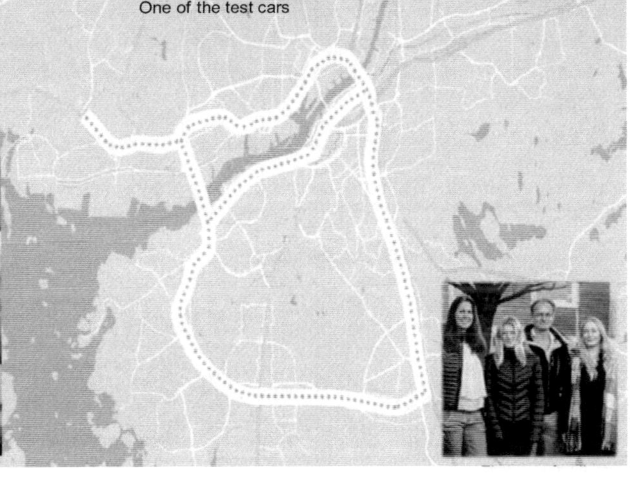

TRAFIKVERKET
SWEDISH TRANSPORT ADMINISTRATION
CHALMERS
VOLVO
City of
Gothenburg
LINDHOLMEN
SCIENCE PARK
Autoliv
Drive Me
SELF-DRIVING CARS FOR
SUSTAINABLE MOBILITY
YFC 017

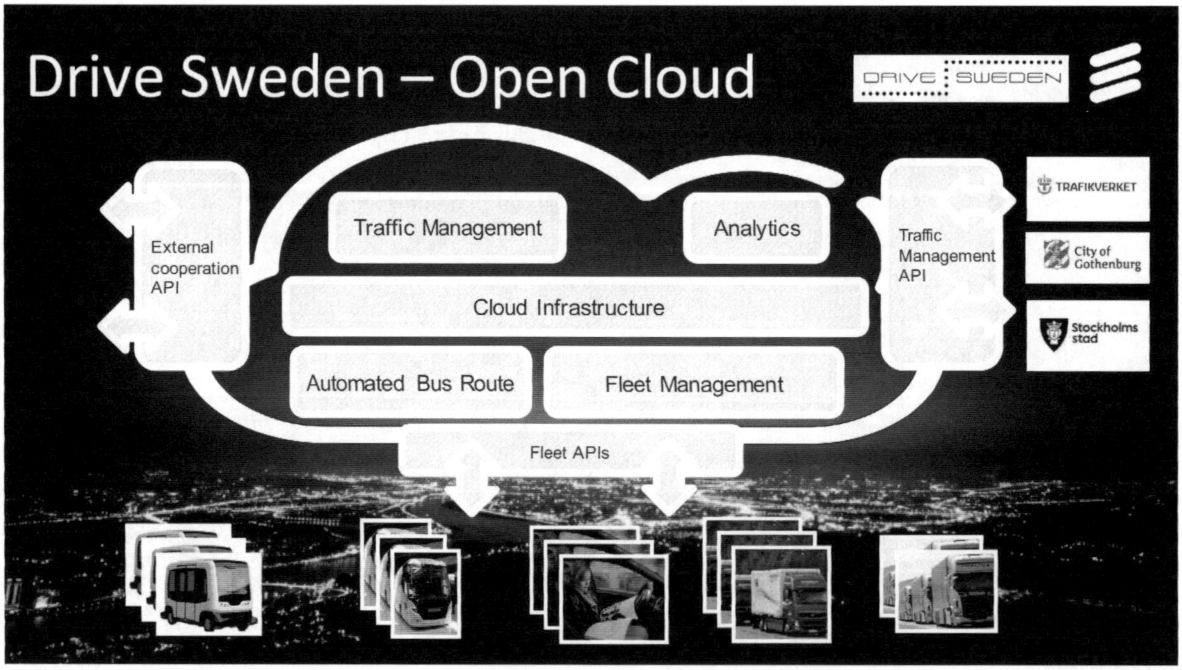
Drive Sweden – Open Cloud
DRIVE SWEDEN
External cooperation API
Traffic Management
Analytics
Cloud Infrastructure
Automated Bus Route
Fleet Management
Traffic Management API
TRAFIKVERKET
City of Gothenburg
Stockholms stad
Fleet APIs

AD Aware Traffic Control

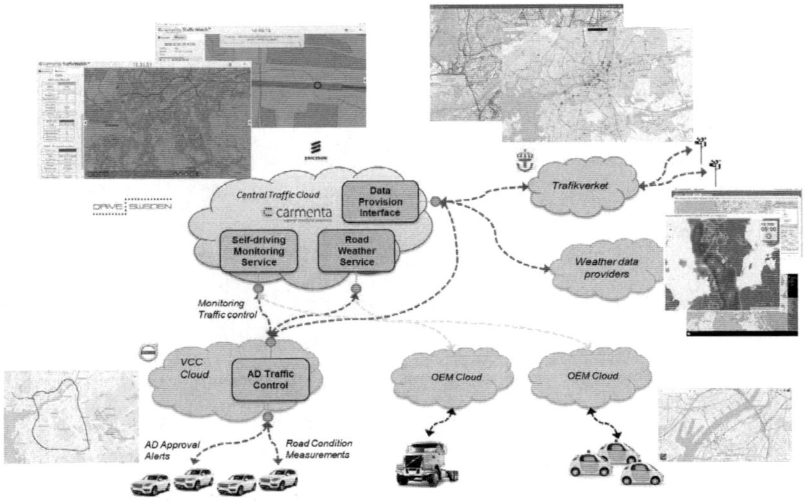

The feature will be used to enable/disable self-driving of the Drive Me cars, individually per road segment, based on real-time weather and incident data.

Now under development:

Include information about emergency vehicles coming from behind

DRIVE SWEDEN

ERICSSON carmenta superior situational awareness TRAFIKVERKET Göteborgs Stad

ASTAZERO

Area: 2000 acres

Rural road: 5,7 km
City area: 4 blocks
High speed: 45,000 m^2
Multi-lane: 700 m

Dedicated 4G/LTE cellular network

5G network now under construction

THE WORLD'S FIRST FULL-SCALE TEST FACILITY FOR TOMORROW'S ACTIVE SAFETY.

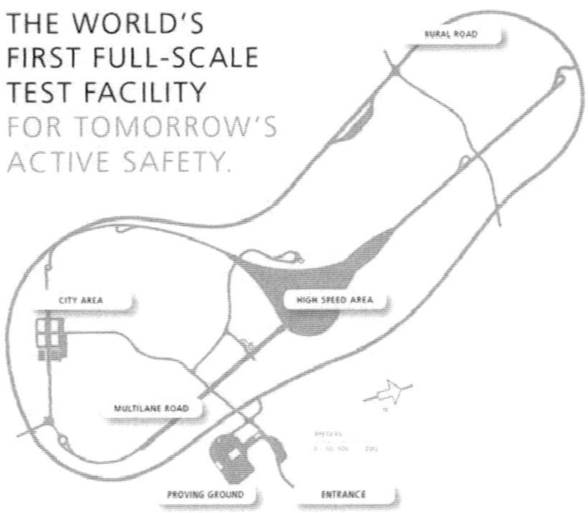

Living labs launched by Sweden's PM

- The Swedish Government has launched a new program, promoting deployment of "living labs" for a wide range of new technologies
- Drive Sweden has been appointed to organize two of these

DRIVE SWEDEN

FRAMEWORK FOR MOBILITY-AS-A-SERVICE IN SWEDEN

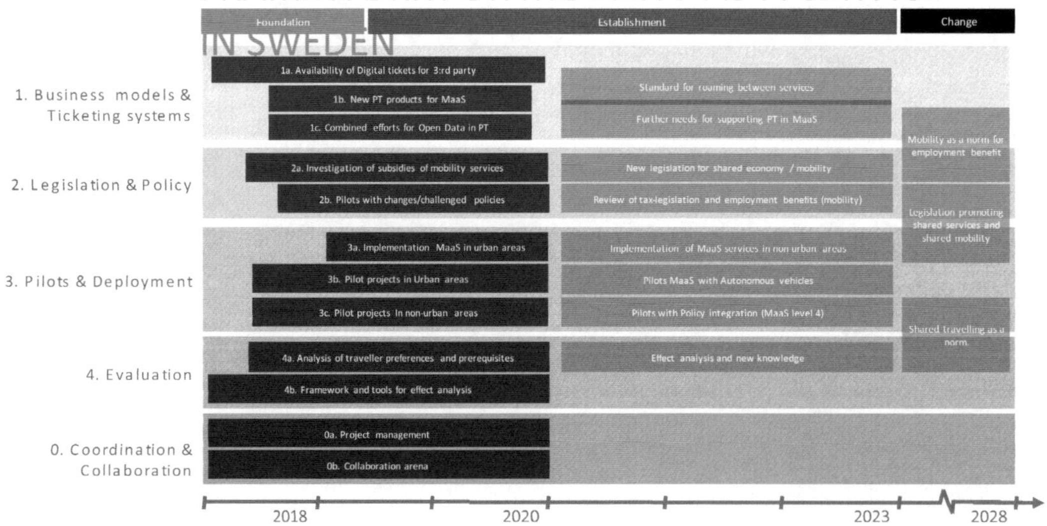

DRIVE SWEDEN

Autopilot, Stockholm

Automated docking & 'bustrains', Gothenburg

Shared Shuttle Services, Gothenburg

Cloud-connected traffic signals, Stockholm & Gothenburg

Simplifying car-sharing, Volvo Cars

Regulatory sand-box

DRIVE SWEDEN

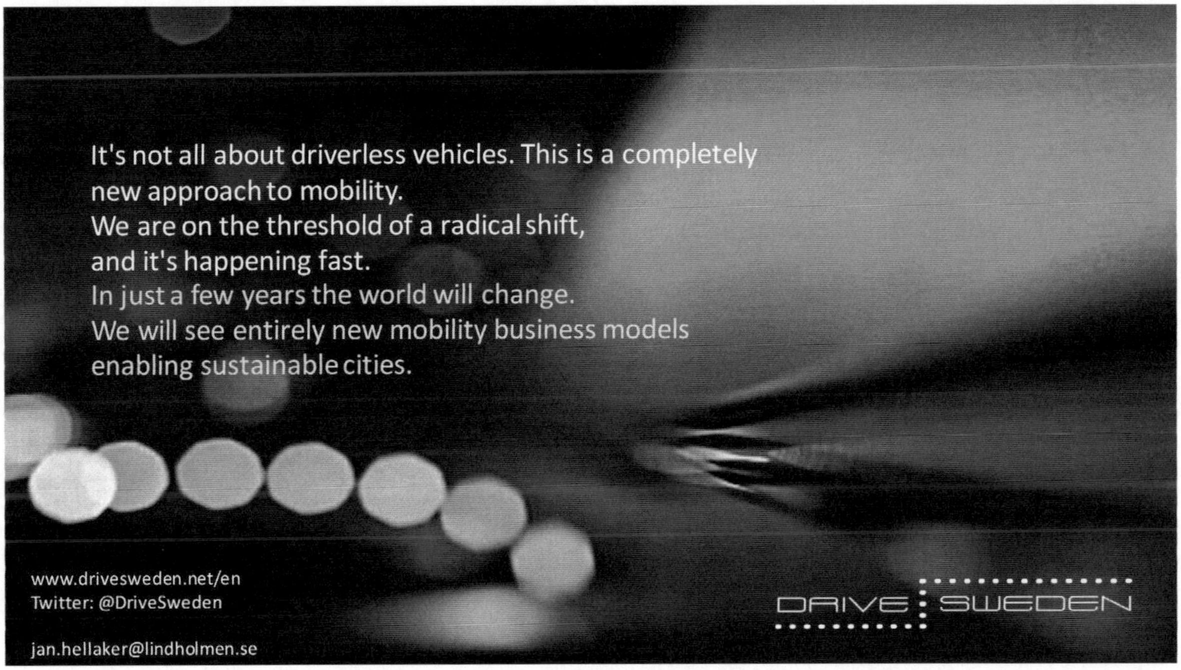

Korea's Automated Vehicle Policies

July, 2017

Ministry of Land,
Infrastructure and Transport

1

Contents

2

1. Conditions and achievement of Korea

3

1-1. Korean conditions for developing automated vehicle

☐ Vehicle production in 2016 was 4.23 million, ranked 6th in the world

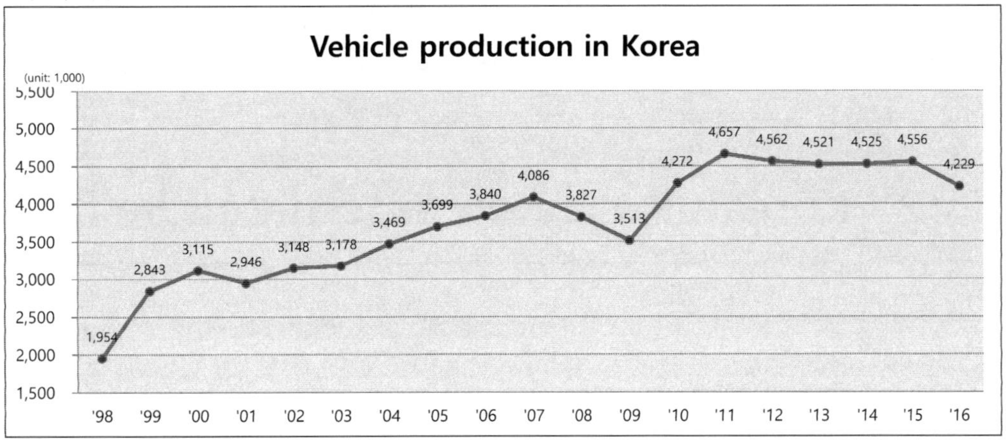

☐ World's best IT/communications infrastructure is built nationwide. Not only car manufacturers like Hyundai, but also global corporations like Samsung are developing automated vehicle

※ Penetration rate of broadband internet: 98%, it average speed: 28.6Mbps / super high speed mobile service ranked first among OECD member economies

4

1-2. Major achievement

Test drive

- Permit test drive of automated vehicle on every road in the country (2016)
- Permit test drive of 'No Steering No Pedal' automated vehicle (2017)
- Public demonstration of automated driving at the central area of Seoul (2017)

Infrastructure

- Began free service of 3D precision map built by the government (2016)
- Began C-ITS trial service (2016, 88km)

Technical support

- Began R&D on safety standard and construction of testbed 'K-City' (2016)
- Create a public-private experts' consultative body (2016)

5

2. Korean Policy on automated vehicle

6

Support policy on commercializing automated vehicle

Vision: Improve transport safety & create a new growth engine

Goal:
- Commercialize Level 3 automated vehicle by 2020
- Lay the foundation for Level 4 - 5 full automated driving era

Experiment Foundation
- Testbed 'K-City'
- Test road, Test driving complex

Pilot project
- Automated driving demonstration at Pyeongchang Winter Olympic Games
- Urban driverless shuttle

Infrastructure
- High-definition map
- High-definition GPS
- C-ITS

Improve system
- Temporary driving permission
- Safety standard
- Insurance, inspection
- Public-private consultative group

7

2-1. Lay foundation for experiment (1)

Testbed 'K-City'

- 360,000m[2] testbed constructed by the government
- Create various situations similiar with real road and city conditions
  - highway, suburbs, city center, community, parking facility
- Create real-life like situations for developing automated driving technology and improving safety
- Phased operation: High speed road (2017) → Entire facility (2018)

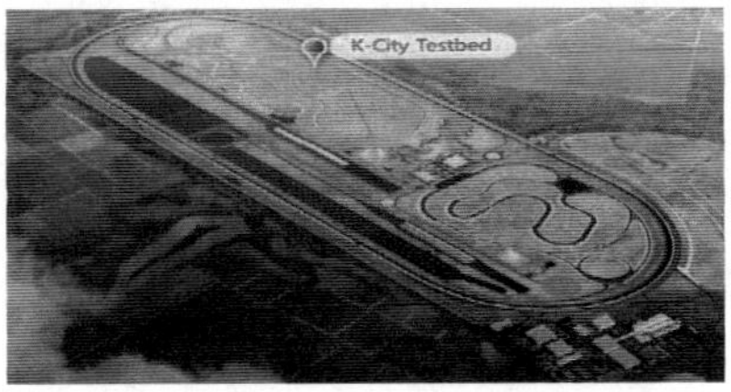

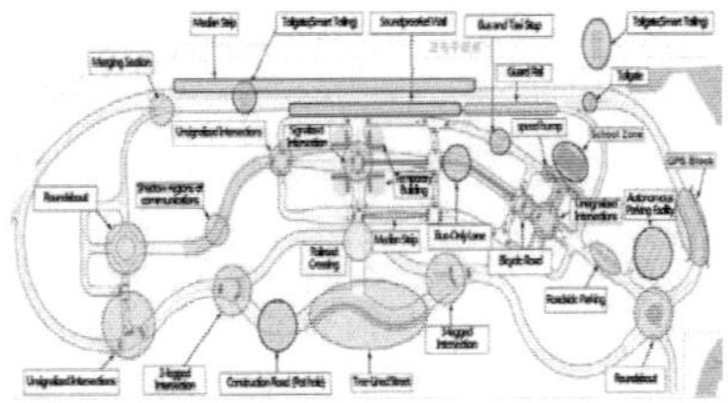

8

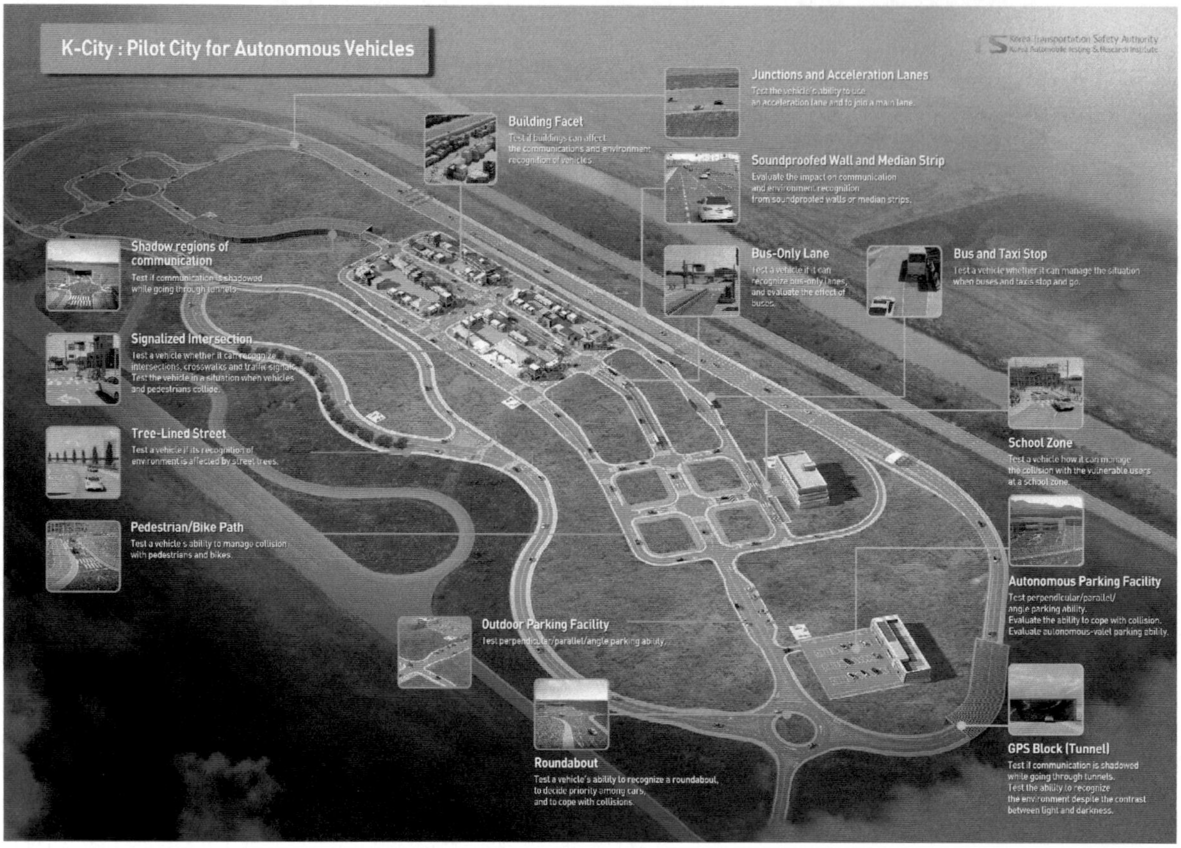

9

2-1. Lay foundation for experiment (2)

Build test road

- Open a 7.7km closed circuit for automated driving (2016)
- Build infrastructure for automated driving such as High-definition map and C-ITS (June, 2017)
- Requests will be accepted, including adjusting the brightness of lanes, intentionally damaging the road, and changing the location of signs

Designate areas for test operation

- Designate advanced industrial complexes as "automated vehicle test operation area" (2016, Pangyo and Daegu)
- Build infrastructure for supporting automated vehicle and conduct pilot projects

10

2-2. Conduct pilot projects

Demonstrate automated driving at 2018 Pyeongchang winter Olympic Games

- Automated driving from Seoul to Pyeongchang for 200km on highway (February, 2018)
  - Include tollgate, intersection, junction and tunnel
- Shuttle service around Olympic stadium (February, 2018)

Operate driverless Shuttle in a city

- Driverless shuttle service (December, 2017)
- Drive side by side with conventional cars for 5 km long at highly populated city near Seoul

11

2-3. Build infrastructure (1)

High-definition 3D map (scale: 1/500)

- Provide information through on/offline channels, including 11 road information
  - lane (4 indicators), road facility (4 indicators), signal (3 indicators)
- Expansion: 1,305km (2017) → about 5000km long highways and major city centers across the country (2020)
- Provide precise map to public for free and support the development of automated driving

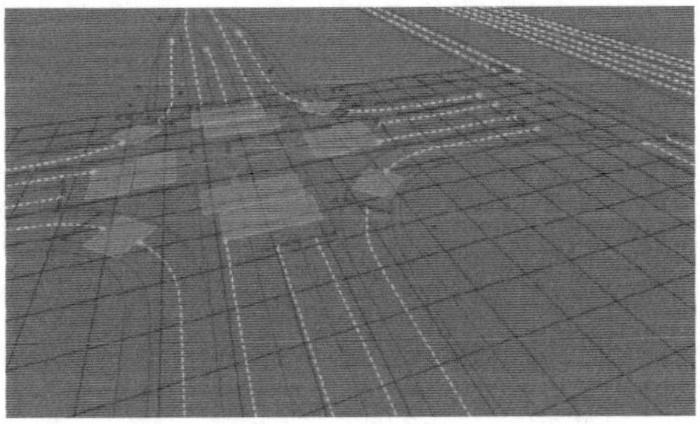

12

2-3. Build infrastructure (2)

High-definition GPS: margin of error is within 1m

- Low cost (using DMB radio wave), quick installation (terminal inside a car)
- Pilot project (2017) → nationwide service (2019)
- With high-definition map, contribute to safer automated driving

13

2-3. Build infrastructure (3)

Cutting edge road system C-ITS

- Real-time traffic monitoring through V2I communication
- Expand the range of sensor and increase installation of sensors
- Introduction: Daejeon - Sejong 88km(2016) → major highways across the country (2020)

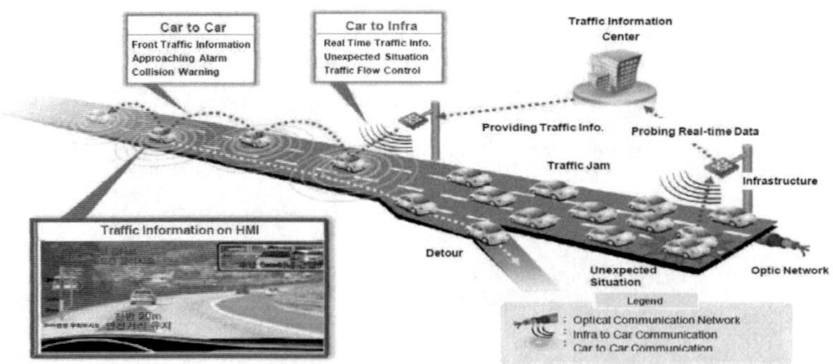

14

2-4. Improve system (1)

Permission for test-driving of automated vehicle

- Adopt permission system for automated vehicle test drive (Feb. 2016)
  - Driver should be on board, put automated vehicle sign, buy an insurance, and pass a pre-test drive (e.g. lane keeping, obstacle avoidance)
  - Request for test drive should be reviewed and responded in 20 days

- Expand test drive area to all roads across the country (Nov. 2016)
  - except for school zones and silver zones

- Exceptional test drive such as 'No Steering No Pedal' are allowed (Mar. 2017)

- So far, 20 cars gained permission
  - including Hyundai motors, Hyundai Mobis, Samsung Electronics, LG Electronics and Seoul Univ.

15

2-4. Improve system (2)

Introduce safety standard

- Created based on safety assessment technology R&D and international standards
  - Develop automated vehicle safety assessment technology and testbed (2016-2018)
  - Develop safety assessment technology for the shift of automated vehicle control, and conduct social acceptability study (2017-2020)

- UNECE/WP.29 activities and harmonization with international standard
  - Actively participate in ACSF and ITS-AD Informal Working Group

Create insurance, inspection and recall system

- Research on insurance for automated vehicle, prepare recall and inspection system for follow-up service

16

2-4. Improve system (3)

☐ Created a public-private consultative body for collecting experts' opinion from every walk of life (June, 2016)

→ About 100 experts are having in-depth discussion on system, technology, society and business.

Automated vehicle convergence future forum

System

- Insurance and legal responsibility in case of accidents
- Direction of amending related laws(road transport, cars)
- Defect investigation and inspection

Technology

- Roads in automated driving era (lane, sign)
- Creating High-definition map and GPS
- Traffic signal system

Society

- Forecast the change of future society
- Draw social consensus (protection of personal information, ethics)

Business

- Nurture new future industries such as share economy
- Expand the added value of data-sharing
- Vitalize automated driving industry and create jobs

17

Future of Korean driven by automated vehicle

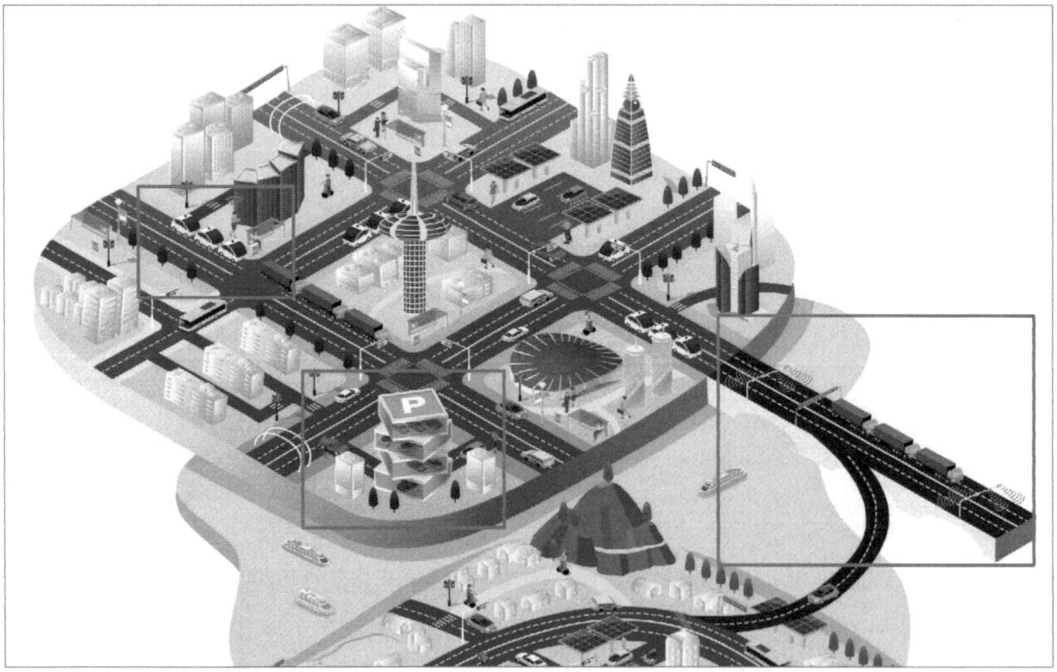

18

Thank you

19

Centre for Connected
& Autonomous Vehicles

Automated vehicles in the UK

Iain Forbes

July 2017

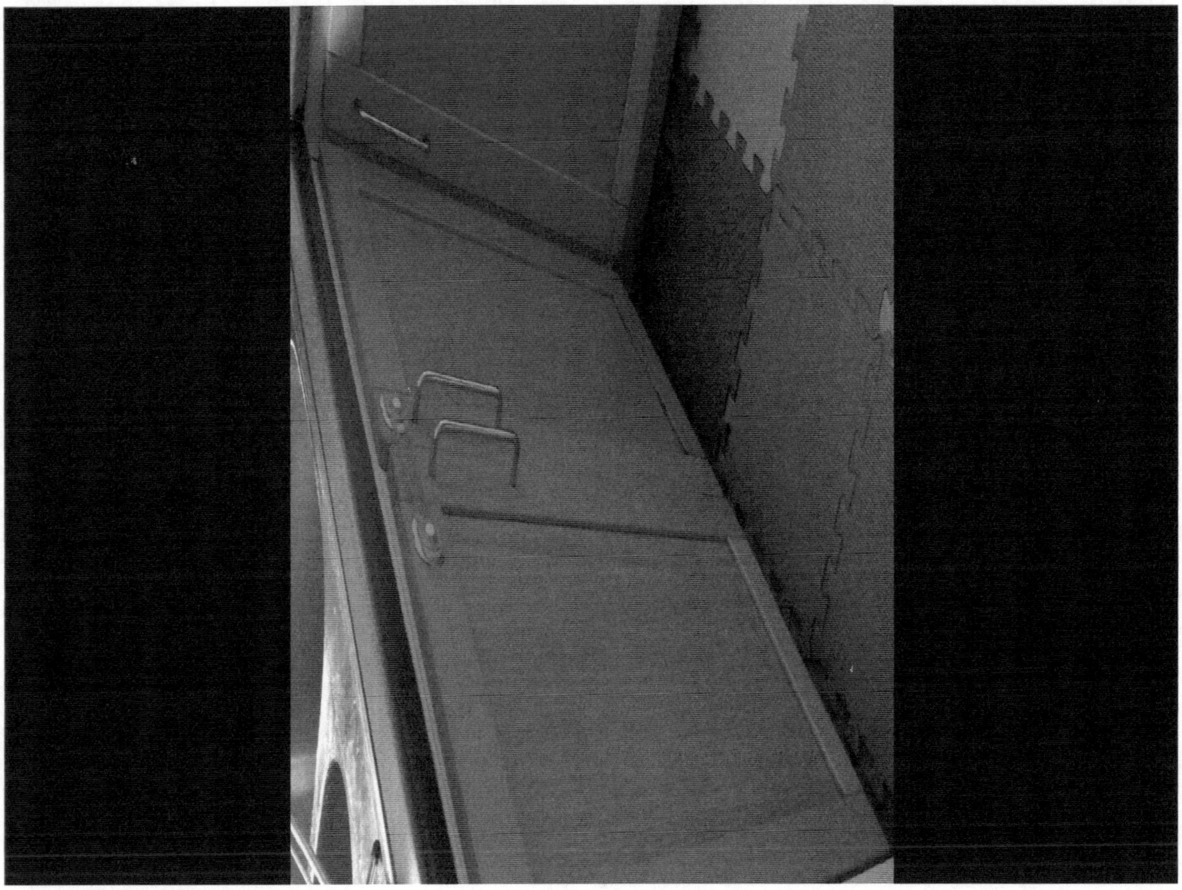

Centre for Connected
& Autonomous Vehicles

Centre for Connected
& Autonomous Vehicles

- Insurance
- Driver training, testing and licensing
- Road infrastructure
- Vehicle roadworthiness and maintenance
- Theft and cyber security
- Other road users
- Active traffic management
- New business models
- Use of the footway
- Mapping
- Vehicle tax, registration & licensing
- Connectivity
- National and international rules on safe use of automated vehicles
- Criminal and civil liability
- National and international automated vehicle standards
- Automated passenger and goods services
- Accident investigation
- Product liability
- Data
- Land use policy

Centre for Connected
& Autonomous Vehicles

Centre for Connected
& Autonomous Vehicles

Centre for Connected
& Autonomous Vehicles

Centre for Connected
& Autonomous Vehicles

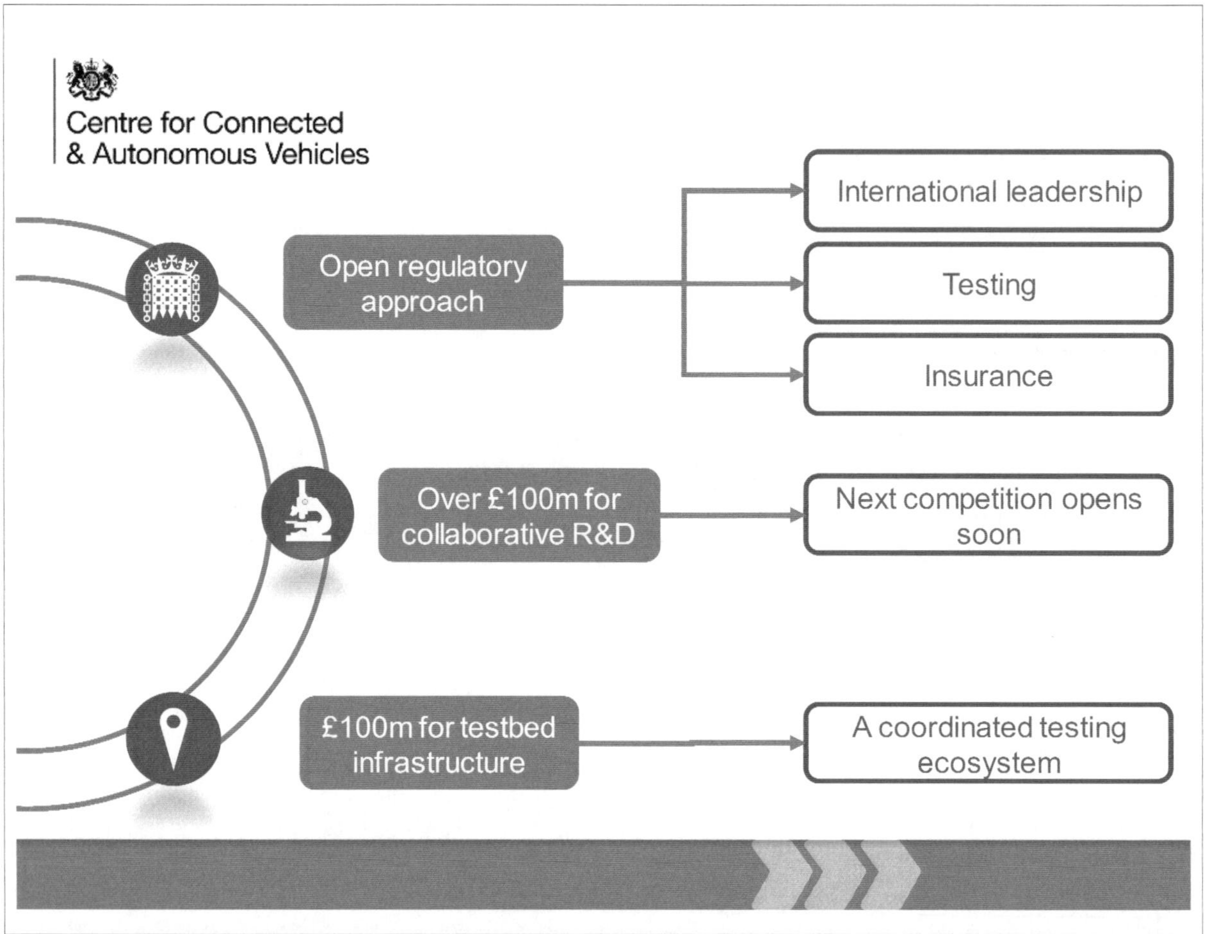
Centre for Connected
& Autonomous Vehicles
Open regulatory approach
International leadership
Testing
Insurance
Over £100m for collaborative R&D
Next competition opens soon
£100m for testbed infrastructure
A coordinated testing ecosystem

@ccavgovuk

Automated Vehicles Symposium 2017

Cross-ministerial Strategic Innovation Promotion Program

SIP Automated driving systems

— Mobility bringing everyone a smile —

July 13, 2017
Sub Program Director of SIP-adus
Yoichi SUGIMOTO

Outline of SIP

- **Intensive R&D program**
  - ✓ promote 5-years R&D (FY2014 - FY2018)
  - ✓ enhancing **cross-ministerial cooperation**
- **11 research themes**
  From societal issues such as Energy, Next-Generation Infrastructures and Local Resources, including R&D for AD
- **Leadership and total Budget**
  CSTI appointed Program Directors and allocates the budget for each research theme. *

* \50bil in total per year
(65% for SIP 11 themes, 35% for medical R&D)

adus : Automated driving systems for universal service

Cross-ministerial Strategic Innovation Promotion Program

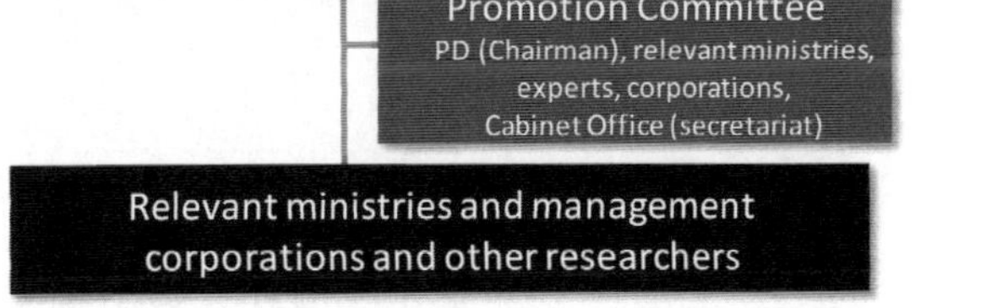

2

Promotion framework of Japanese Government

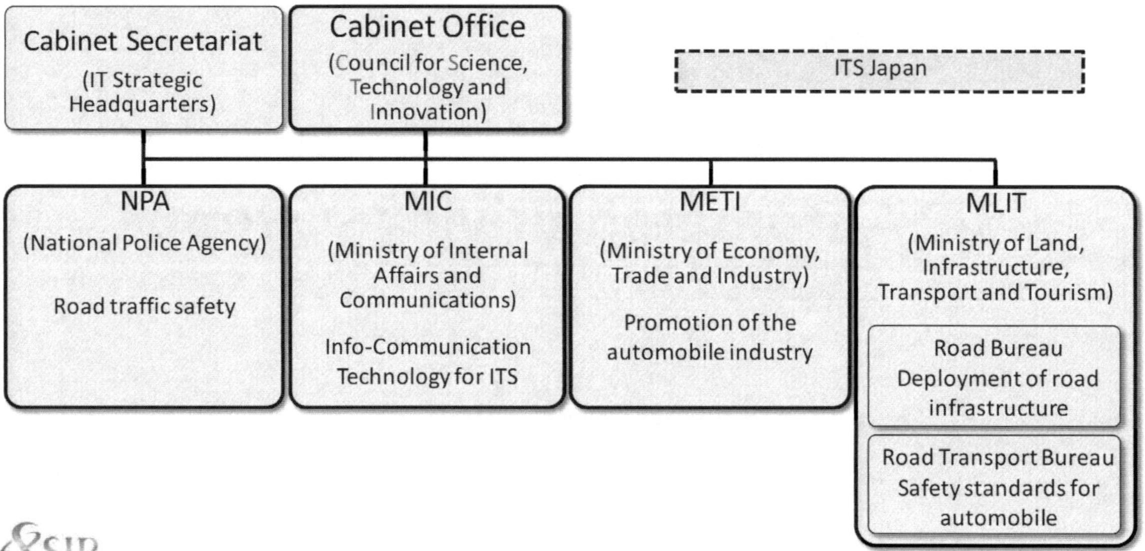

SIP

3

Goal & Exit Strategy

1. Ensuring safety and traffic jam reduction on the road
2. Realization and spread of Automated Driving System
3. Realization of advanced next generation public bus service for vulnerable people.

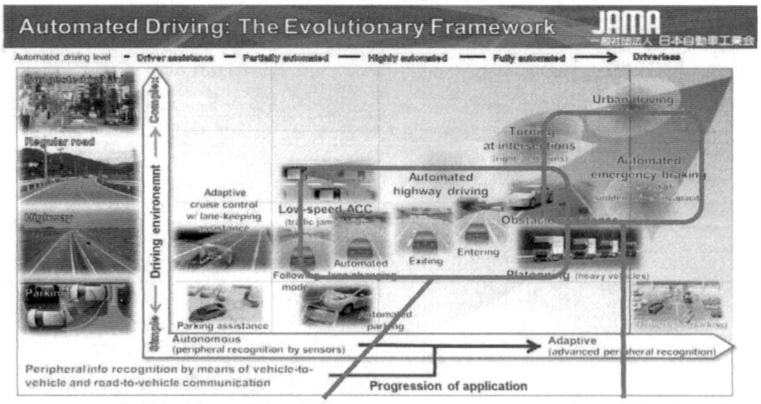

Realization of Level 2 on highway by 2020

Prioritization for next step Level 2 on regular road

SIP

4

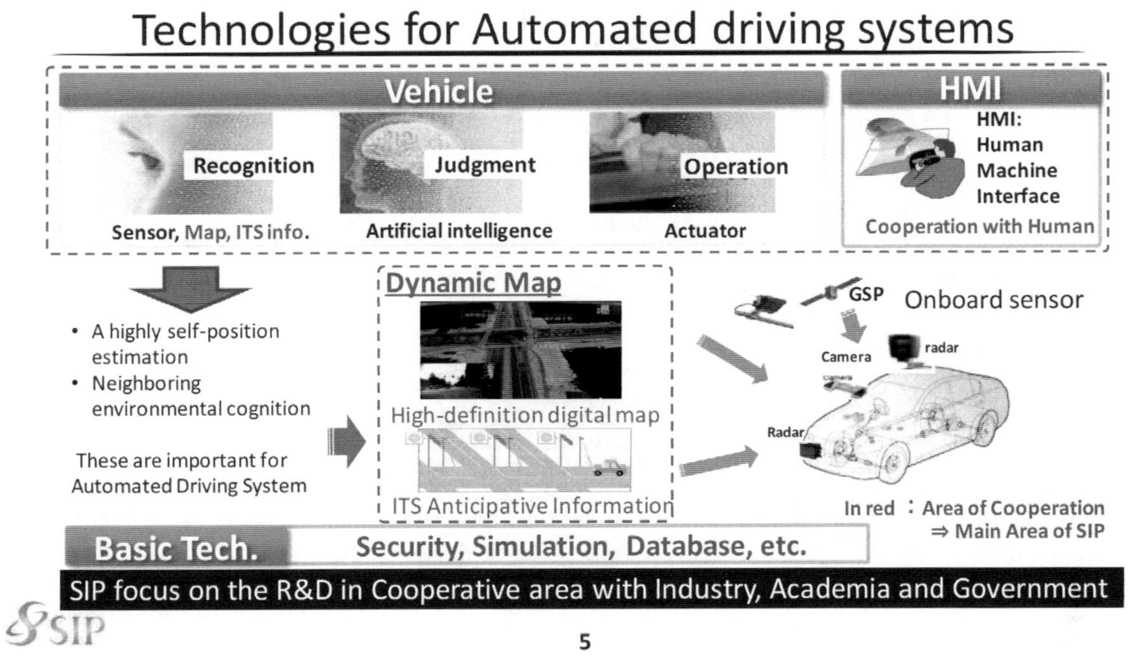

Technologies for Automated driving systems
Vehicle
HMI
Recognition
Judgment
Operation
HMI:
Human
Machine
Interface
Sensor, Map, ITS info.
Artificial intelligence
Actuator
Cooperation with Human
Dynamic Map
GSP
Onboard sensor
A highly self-position estimation
Neighboring environmental cognition
These are important for Automated Driving System
Camera
radar
High-definition digital map
Radar
ITS Anticipative Information
In red : Area of Cooperation
⇒ Main Area of SIP
Basic Tech.
Security, Simulation, Database, etc.
SIP focus on the R&D in Cooperative area with Industry, Academia and Government
SIP
5

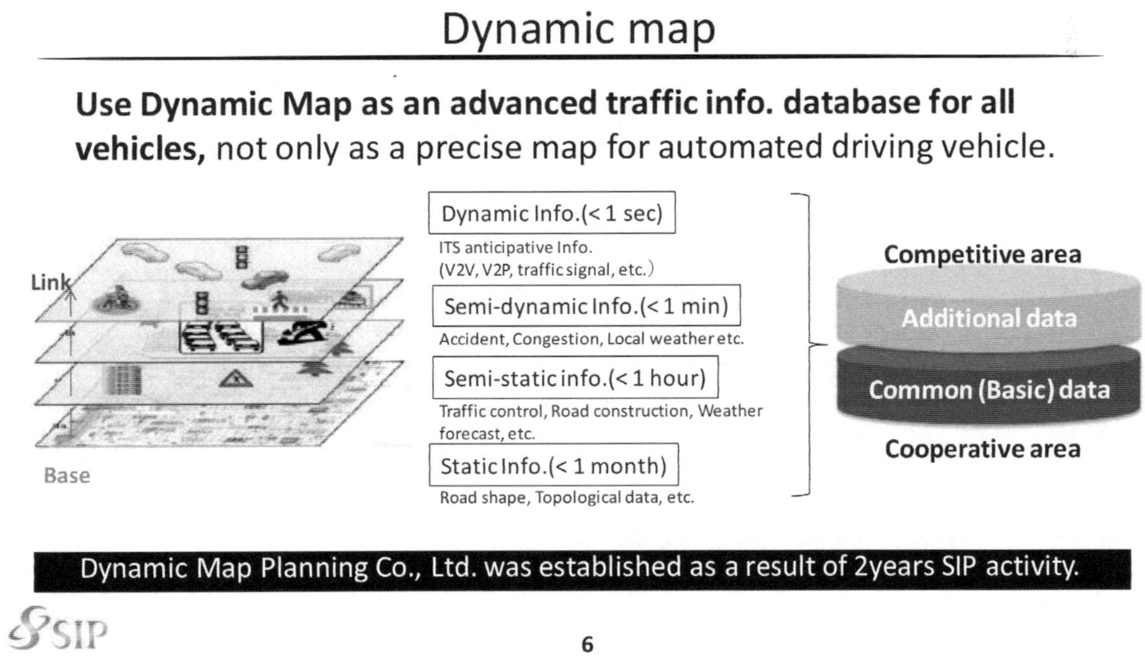

Dynamic map
Use Dynamic Map as an advanced traffic info. database for all vehicles, not only as a precise map for automated driving vehicle.
Link
Base
Dynamic Info.(< 1 sec)
ITS anticipative Info.
(V2V, V2P, traffic signal, etc.)
Semi-dynamic Info.(< 1 min)
Accident, Congestion, Local weather etc.
Semi-static info.(< 1 hour)
Traffic control, Road construction, Weather forecast, etc.
Static Info.(< 1 month)
Road shape, Topological data, etc.
Competitive area
Additional data
Common (Basic) data
Cooperative area
Dynamic Map Planning Co., Ltd. was established as a result of 2years SIP activity.
SIP
6

Establishment of Dynamic map database

Dynamic map

【Required condition】
- Freshness of data/Easiness of data updating
- Scalability
- Low cost
- Security etc.

- **Data commoditizing**
- **Utilization of probe data**

Public-Private Partnership

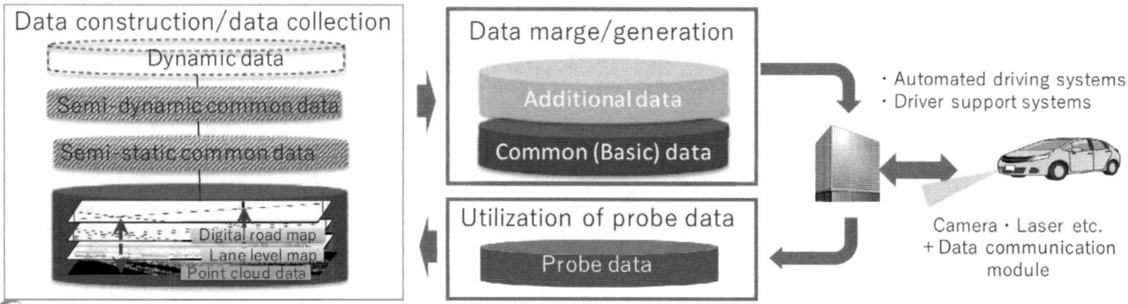

SIP

7

Dynamic Map Planning Co., Ltd

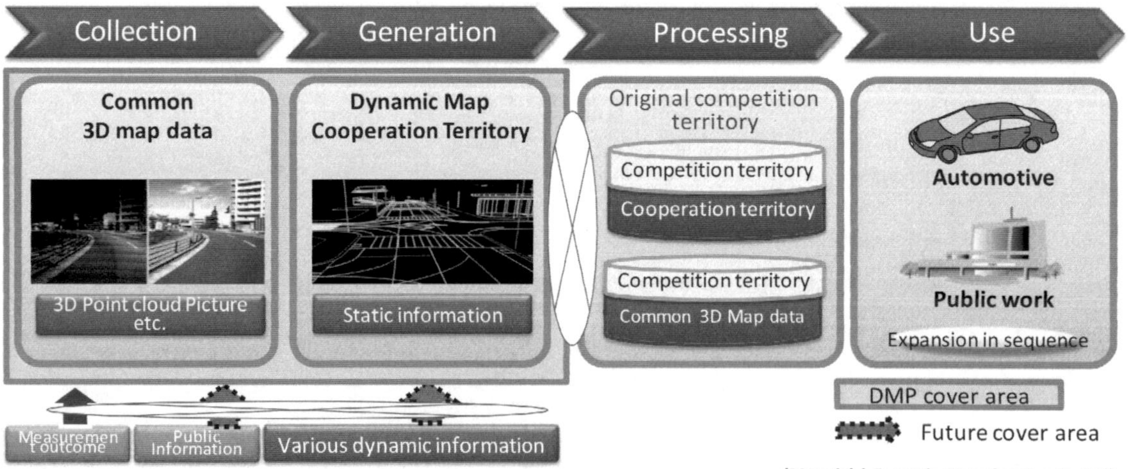

(Material：Dynamic Map Planning Co. Ltd)

SIP

8

Cyber security

①Common Models of AD for Threat Analysis
②Validation/Evaluation Methods and Criteria
③Certificate Validation of V2X communication

(※For specific study, **closely collaborate with SIP cyber security**)

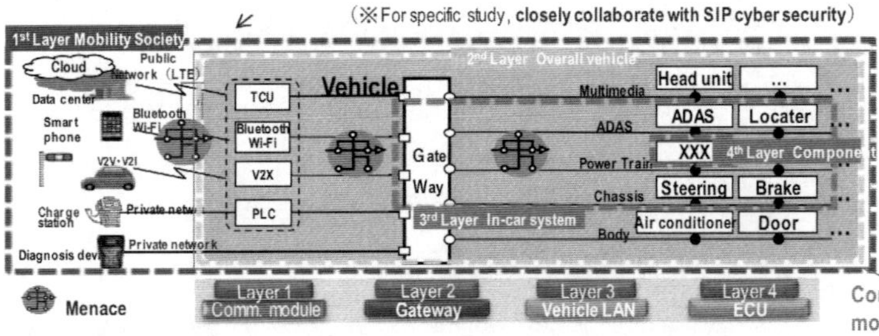

Common models

①To establish threat analysis

①Common architecture model
①Use cases of AD (JAMA)
②Threat Info. (JPCERT/CC、Auto-ISAC)
②Evaluation（Attack）Info (Auto-ISAC)

①User friendliness (JAMA)

①Threat analysis tool

①Comparison with current threat analysis (Jaspar)

②To establish Test protocols

②Countermeasure
②Level of Countermeasure

③To reduce V2X signature

SIP

9

Human Machine Interface

1) To investigate effects of system information on drivers' behavior.
2) To investigate effects of driver state on his/her behavior in transition.
3) To investigate effective ways to functionalize AV to be communicative

Driver state

- Cognitively distracted
- Physically distracted
- Low arousal
- Lack of SA
- Out of position

Controlled

Readiness

- Head orientation and visual performance
- Heart rate and blood pressure
- Body temperature
- Skin conductance
- EEG
- Posture and body

Performance at the event

- Longitudinal and lateral control of the vehicle
- Minimum distance and minimum TTC to the hazard
- Time spent to regain control

Correlation

SIP

10

Pedestrian collision reduction

Vehicle-to-Pedestrian (V2P) Communication

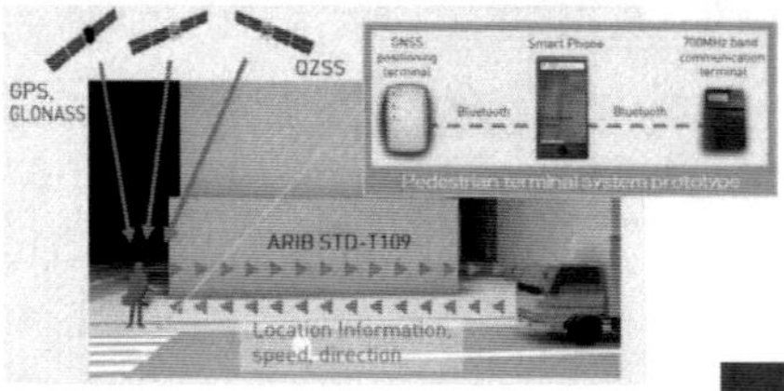

700MHz Direct Wireless Communication

- 700MHz band communication
- High-precision positioning
- Danger identification and pedestrian safety support

Infrastructure radar with V2I communication

79GHz band radar from roadside of intersection

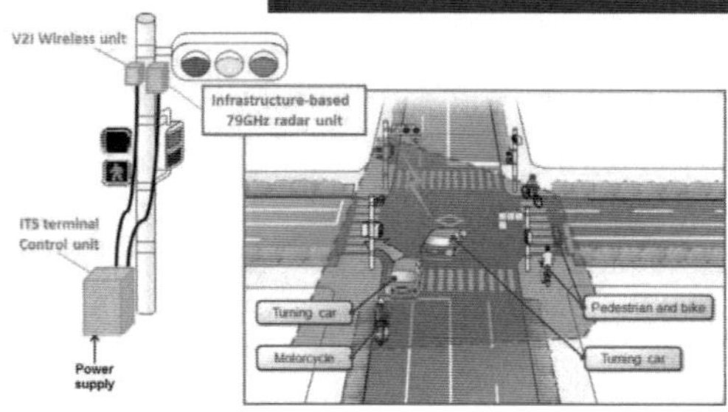

SIP

Next generation Transport

ART information center

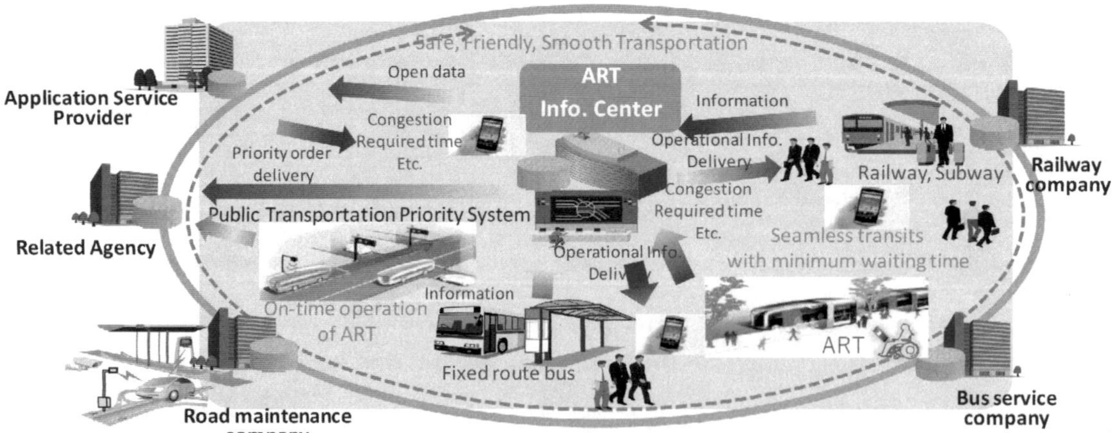

Information Open Platform for ART related applications

Application Examples
* Congestion Prediction
* Dynamic Connection Guidance
* Remote Diagnostics

SIP

Field Operational Test (FOT)

≪Purpose≫

1. To activate the R&D
2. To prove each elemental technology
3. To enhance international cooperation and harmonization
4. To Build Social acceptance

≪Participant≫

- OEM/Supplier
- University/Research organization
- Ministries, government officers
- **Foreign OEM/supplier**
- Journalist

≪Period≫

Autumn 2017 ～ beginning of 2019

≪Main themes≫

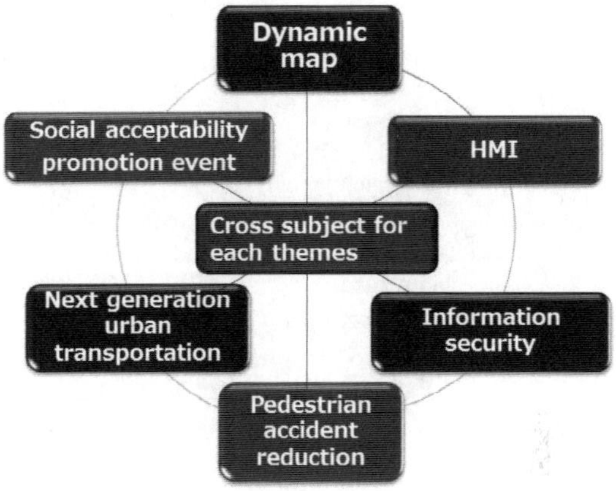

SIP

13

Field Operational Test (FOT)

≪Test site≫

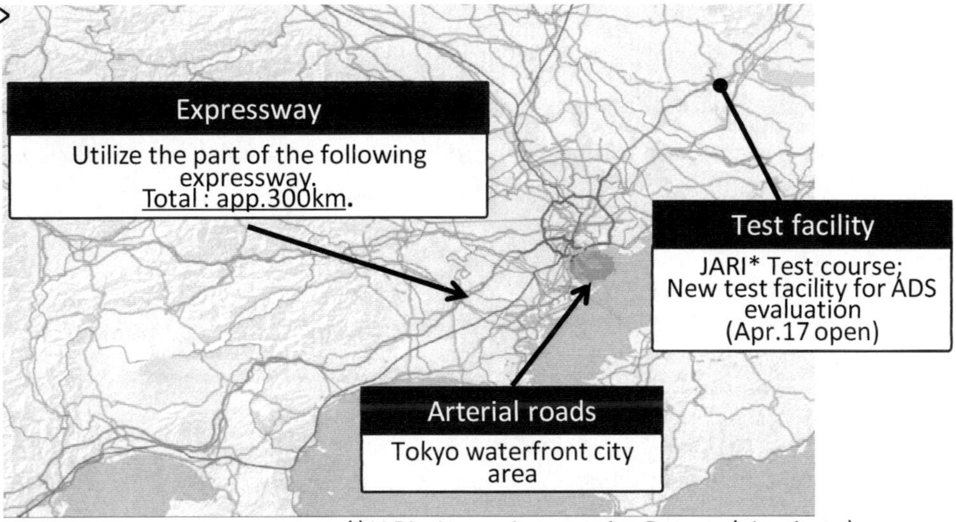

(*JARI : Japan Automotive Research Institute)

SIP

14

Field Operational Test (FOT)

Dynamic Map(Example)

- ☐ To validate 3D high-resolution digital map data
- ☐ To validate data collection and distribution method
- ☐ To verify the utility of semi dynamic information
- ✓ The map data is provided by SIP-adus.

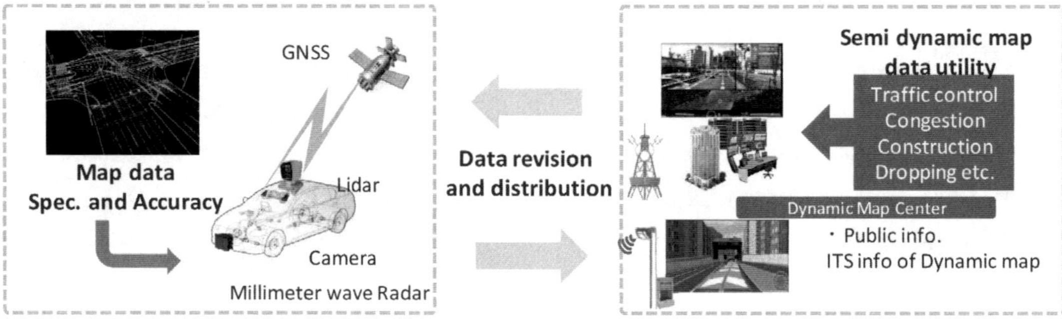

15

Field Operational Test (FOT)

HMI(Example)

- ☐ To collect and analyze the driver state data
- ☐ To define driving readiness status and driver
- ☐ Verification of HMI methods and devices .

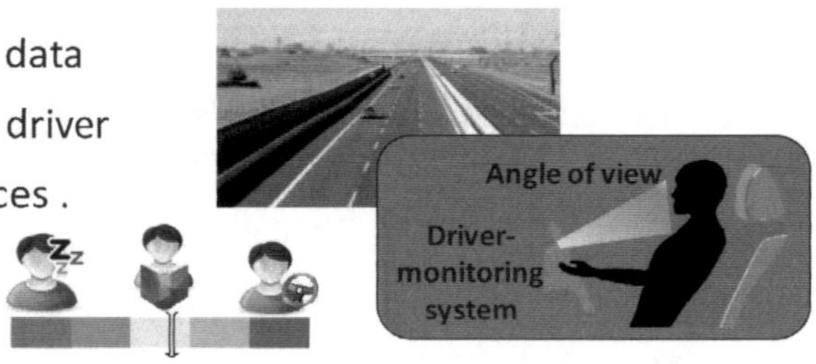

Cyber Security(Example)

- ☐ To Validate the evaluation method
- ☐ Inspect defense functions of ADV

Layer1: Communication of Out Car
Layer2: E/E Architecture
Layer3: In Car Bus Protocol
Layer4: ECU Software Structure

16

Automobile Society

1886 BENZ Patent Motorwagen

(Toyota Automobile Museum)

1907 Piccadilly Circus, London

(Alpshima; sohske.cocolog-nifty.com)

SIP

17

Common Platform

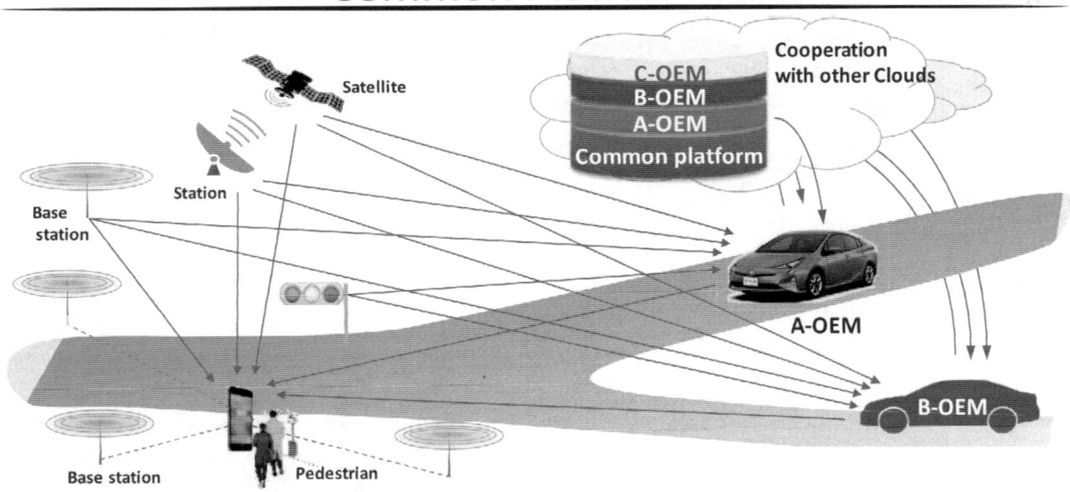

A common base platform(map, rule etc.) is necessary for keeping safety and the social order.

⇒ **Promoting harmonization and standardization**

SIP

18

International Cooperation activities

■ **Experts assigned in Focused areas**

1. Dynamic Map
2. Connected Vehicle
3. Human Factors
4. Impact Assessment
5. Next Generation Transport
6. Security

EU★US★JAPAN
ITS COOPERATION

ISO International Organization for Standardization

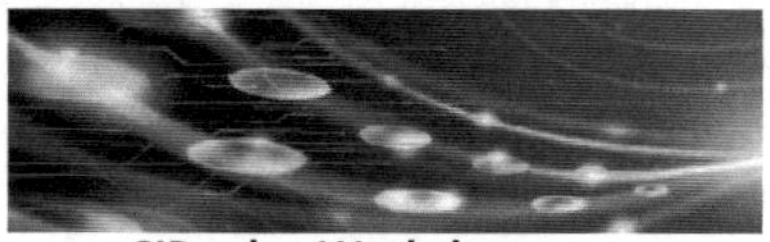

SIP-adus Workshop

SIP

19

SIP-adus Workshop 2016

Recognized as a specialized international conference on automated driving, participants from all over the world increased.
Sharing latest information, building friendship among experts, were highly evaluated by the participants.

- **Organizer** : Cross-Ministerial Strategic Innovation Promotion Program, Council for Science, Technology and Innovation, Cabinet Office, Government of Japan
- **Date** : November 15-17, 2016
- **Venue** : Tokyo International Exchange Center http://www.jasso.go.jp/tiec/index_e.html
- **Attendees** : 425 from 17 countries
- **Speakers** : 61 includes 34 speakers and moderators from overseas

Snapshot with speakers from overseas after Minister Tsuruho

SIP

20

Thank you for kind attention!

SIP

VDI|VDE|IT

AUTOMATED VEHICLE INITIATIVES: **EUROPEAN UNION**

Dr. Gereon Meyer
Head of Strategic Projects
VDI/VDE Innovation + Technik GmbH
Berlin, Germany
gereon.meyer@vdivde-it.de

EUROPEAN COMMISSION

EUROPEAN COMMISSION VDI|VDE|IT

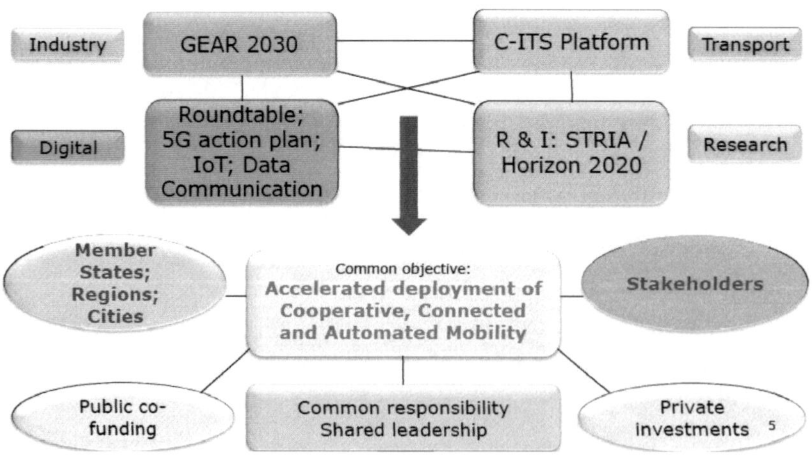

Policy, regulation, and innovation for connected and automated driving

ETP ROADMAPS: ERTRAC

VDI|VDE|IT

AUTOMATED DRIVING ROADMAP

www.ertrac.org

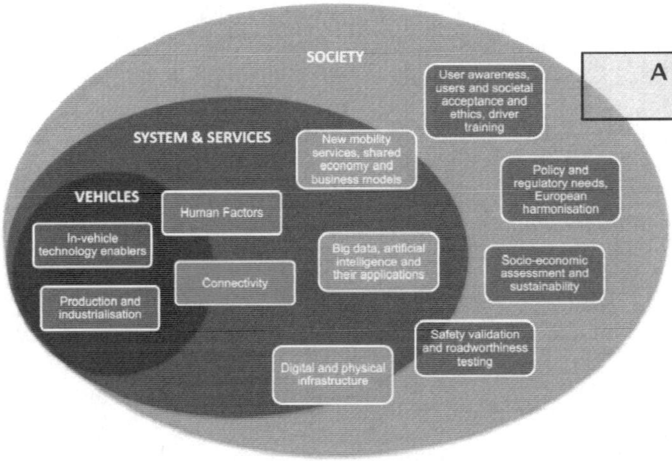

Plans for connected / automated passenger, freight and urban mobility

ETP ROADMAPS: EPoSS

EPoSS European Technology Platform on Smart Systems Integration

VDI|VDE|IT

EUROPEAN ROADMAP
SMART SYSTEMS FOR
AUTOMATED DRIVING

www.smart-systems-integration.eu

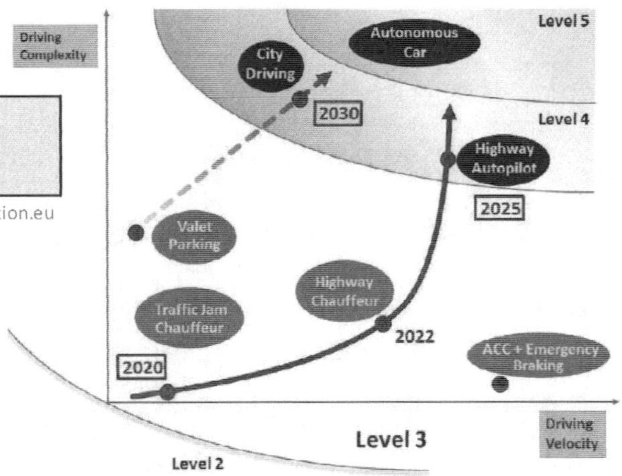

Technology needs for evolutionary and revolutionary developments

EU ROADMAPS

EUROPEAN COMMISSION

VDI | VDE | IT

STRATEGIC TRANSPORT RESEARCH AND INNOVATION AGENDA (STRIA)

European Commission
Staff Working Document (2017), 223

e.g. 2020 Actions for Connected and Automated Road Transport

	Successful technological development	Swift deployment of CAT technologies and increased competitiveness of the European industry	Enabling framework, including new regulations/ legislation, pre-conditions, infrastructure
LARGE-SCALE, CROSS BORDER DEMOS	Large-scale, cross-border demos for freight and passengers: (all modes and multimodal/door-to-door). 1) Develop and test shared, connected and cooperative automated vehicle fleets in urban areas for the mobility of all 2) Develop and test connected and automated heavy-duty vehicles in real logistics operations 3) Test the applicability of 5G connectivity to "Connected and Automated Driving" use cases 4) Large-scale, cross-border demonstration of highly automated driving functions for passenger cars. Demos will analyse different business models/implementation scenarios to meet the demand of different user groups in different regions and operating environments. Demos will consider the optimised use of digital technologies as the Internet of Things, Artificial Intelligence and Big Data for automation.		Knowledge and data platform(s) to enable MSs coordination and exchange of knowledge and data from demos **GEAR 2030 (Road)**: address important legal and regulatory issues and standardisation for connected and automated vehicles **C-ITS Platform(Road)**[30]: recommendations on C-ITS to use the potential for connected and automated vehicles **Round Table on Connected and Automated Driving** (Road) addressing digital issues Actions associated with the **The Letter of Intent on cross-border testing** signed by 29 European countries on 23 March 2017[31]

Comprehensive top-down strategy covering all transport modes

SOCIETAL FACTORS

MOBILITY 4 EU

VDI | VDE | IT

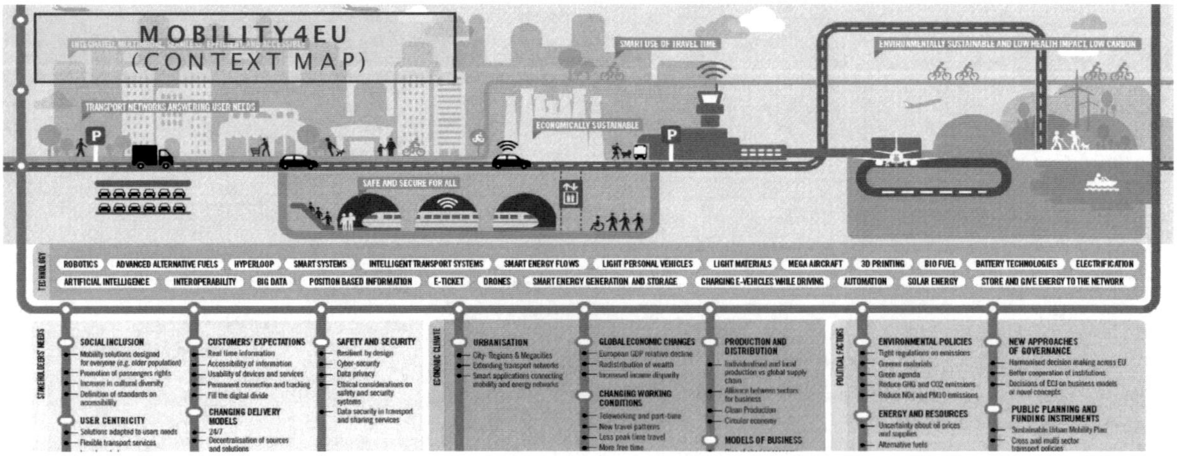

www.mobility4eu.eu

Bottom-up initiative co-creating a multi-modal action plan with users

HORIZON 2020 – ART CALLS 2016/17

EUROPEAN COMMISSION   VDI|VDE|IT

Topic	Title	Action type	Budget (EUR Mio)	
			2016	2017
ART-02	Automation pilots for passenger cars	IA	48M€	
ART-04	Safety and end-user acceptance aspects of road automation in the transition period	RIA		
ART-05	Road infrastructure to support the transition to automation and the coexistence of conventional and automated vehicles on the same network	RIA	13M€	
ART-06	Coordination of activities in support of road automation	CSA	3M4	
ART-01	ICT infrastructure to enable the transition towards road transport automation	IA		50M€
ART-03	Multi-Brand platooning in real traffic conditions	IA		
ART-07	Full-scale demonstration of urban road transport automation	IA		

CSA = Coordination and Support Action; IA = Innovation Action; RIA = Research and Innovation Action

Implementation of roadmap topics; 2018-19 calls to be published soon

EU FUNDED CAD PROJECTS

EUROPEAN COMMISSION   VDI|VDE|IT

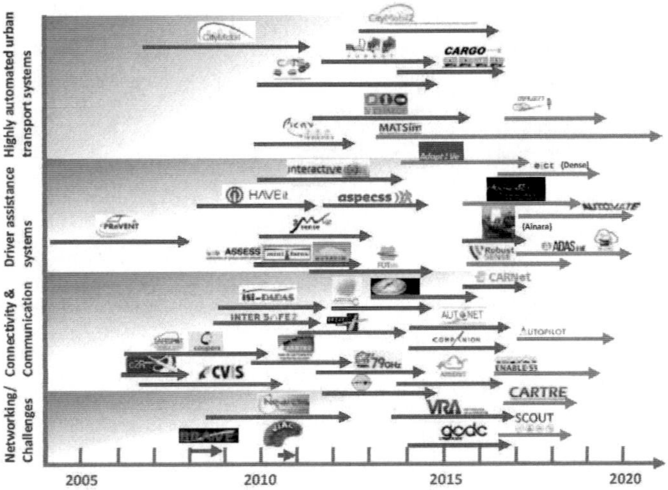

Project Acronym	Project Title	Project Topic Code	Projekt-start
INFRAMIX	Road Infrastructure ready for mixed vehicle traffic flows	ART-05-2016	01.06.17
BRAVE	BRidging gaps for the adoption of Automated VEhicles	ART-04-2016	01.06.17
L3Pilot	Piloting Automated Driving on European Roads	ART-02-2016	1/09/2017
CoEXist	'AV-Ready' transport models and road infrastructure for the coexistence of automated and conventional vehicles	ART-05-2016	01.05.17
Trust Vehicle	Improved trustworthiness and weather-independence of conditional automated vehicles in mixed traffic scenarios	ART-04-2016	01.06.17
TransAID	Transition Areas for Infrastructure-Assisted Driving	ART-05-2016	01.09.17
interACT	Designing cooperative interaction of automated vehicles with other road users in mixed traffic environments	ART-04-2016	01.05.17
CARTRE	Coordination of Automated Road Transport Deployment for Europe	ART-06-2016	01.10.16

Projects of 2016 calls started, 2017 in second application stage

1st EU CAD CONFERENCE

www.connectedautomateddriving.eu

Integrated programme covering EU policy and innovation dimensions

SYNERGIES

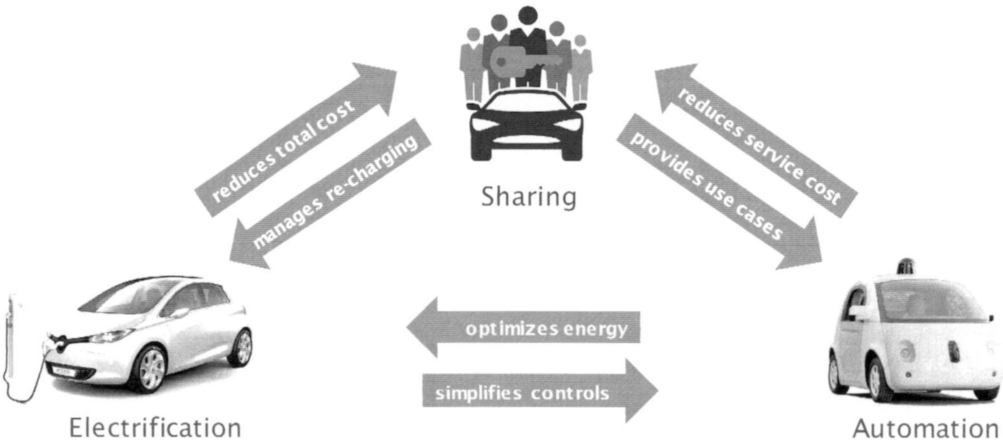

New usage models will tighten the links between the technology paths

BOOK

VDI|VDE|IT

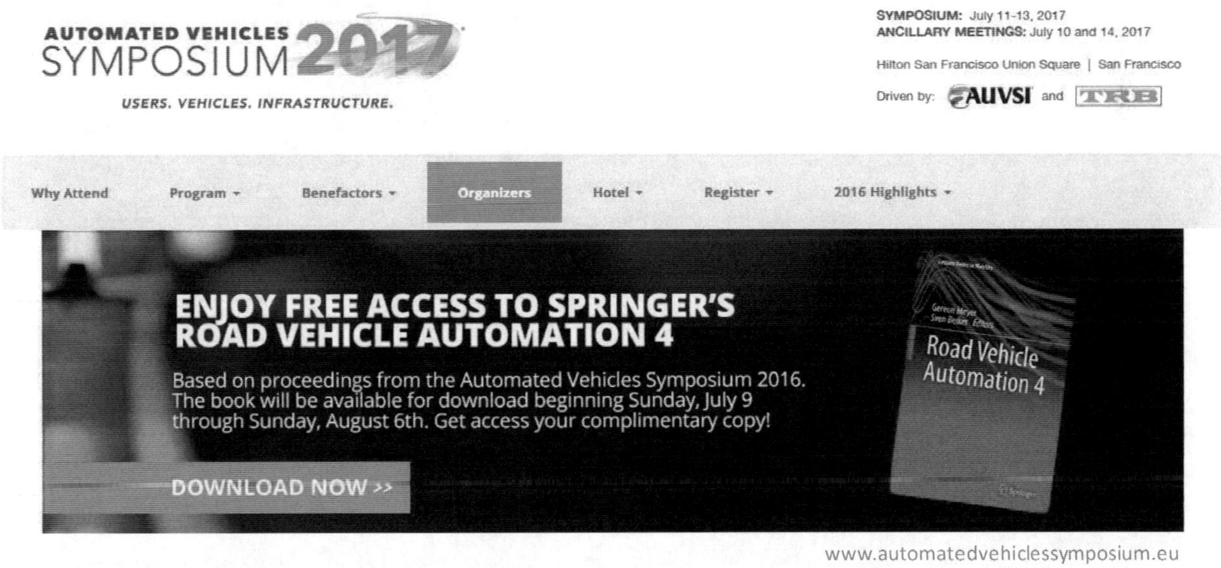

Books available in 300 libraries worldwide; No. 5 in preparation

U.S. DEPARTMENT OF ENERGY

Office of ENERGY EFFICIENCY & RENEWABLE ENERGY

DOE's Focus on Energy Efficient Mobility Systems

David L. Anderson
Energy Efficient Mobility Systems Program
Vehicle Technologies Office

Automated Vehicle Symposium
San Francisco, California
July 13, 2017

MOBILITY IS FOUNDATIONAL TO OUR WAY OF LIFE

Graphics from highwaysindustry.com, mhlnews.com, fhwa.dot.gov, wmata.com, usatoday.com, maritime-connector.com, muffingroup.com, digitaltrends.com, sfmta.com, teslarati.com

MOBILITY IS A LARGE PART OF OUR ENERGY ECONOMY

Transportation is the 2nd largest expense for U.S. households

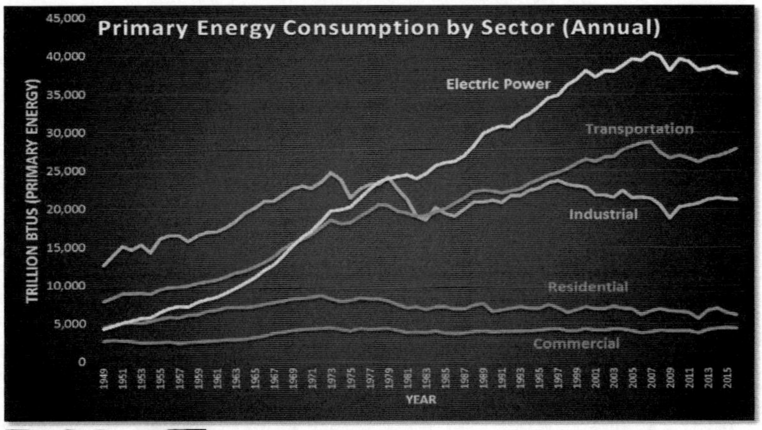

70% of total U.S. petroleum usage is for transportation

On-road vehicles account for **85%** of transportation petroleum usage

CONVERGING TRENDS ARE SHAPING MOBILITY

Population

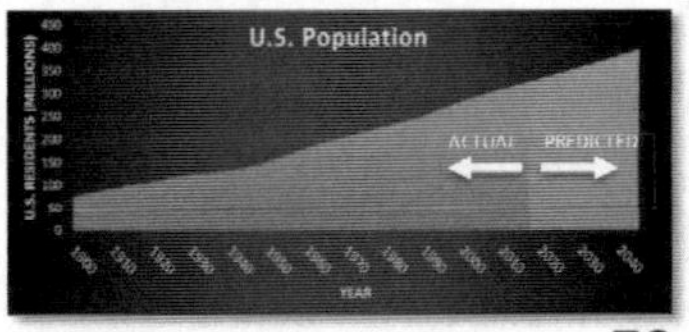

Population expected to grow by **70 million** in next **30 years**

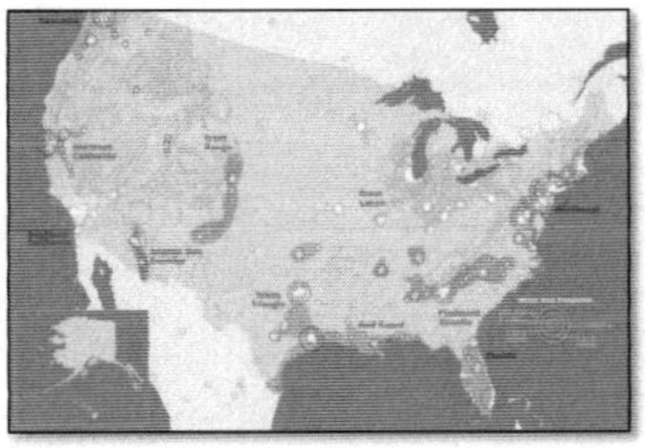

75% of population concentrated in **11** Megaregions

Demographics

Americans are Living Longer

By 2045, the number of Americans over age 65 will increase by **77%**.

About **one-third** have a disability that limits mobility.

Millennials are Connected & Influential

There are **73 million** Americans aged 18 to 34.

They drove **20%** fewer miles in 2010 than at the start of the decade.

Technology

- Integration of Connected & Automated Technologies
- Introduction of Shared Service Platforms
- Advancements in Energy Storage Technology
- Deeper Application of Big Data
- Faster Processing Speeds at Decreasing Cost

TRENDS ARE CAUSING A FUNDAMENTAL DISRUPTION

Connectivity

Ride-hailing

New Powertrains

Industry is leading the introduction of disruptive business models & technologies based on consumer demand.

DOE must understand:

- How will this disruption lead to new ***energy efficiency opportunities?***
- What are the ***risks to energy use*** and how can we overcome them?
- What are the most promising innovation levers for a ***sustainable energy future?***

Automation

Car-sharing

New Modes

FUNDAMENTAL DISRUPTION, DRAMATIC ENERGY IMPACTS

+200%

Potential Increase in Energy Consumption

2050 Baseline Energy Consumption

Potential Decrease in Energy Consumption

-60%

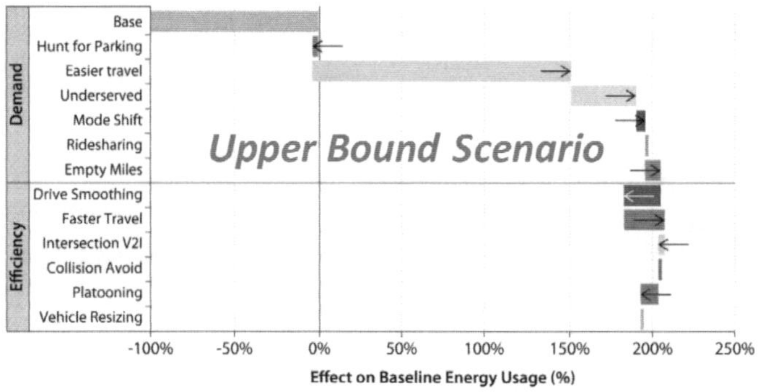

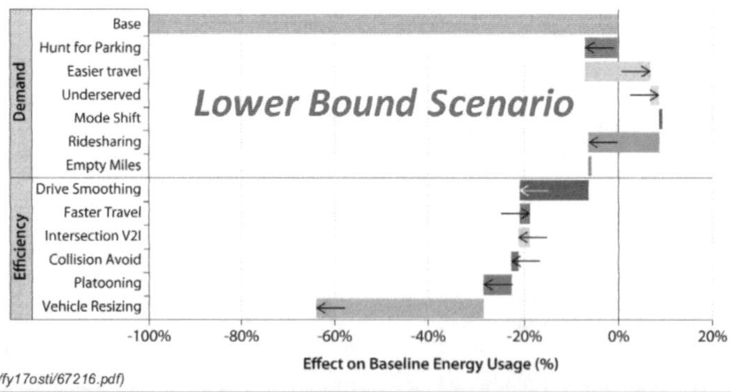

Source: Joint study by NREL, ANL, and ORNL *(http://www.nrel.gov/docs/fy17osti/67216.pdf)*

VTO EXPANDING FOCUS TO TRANSPORTATION LEVEL

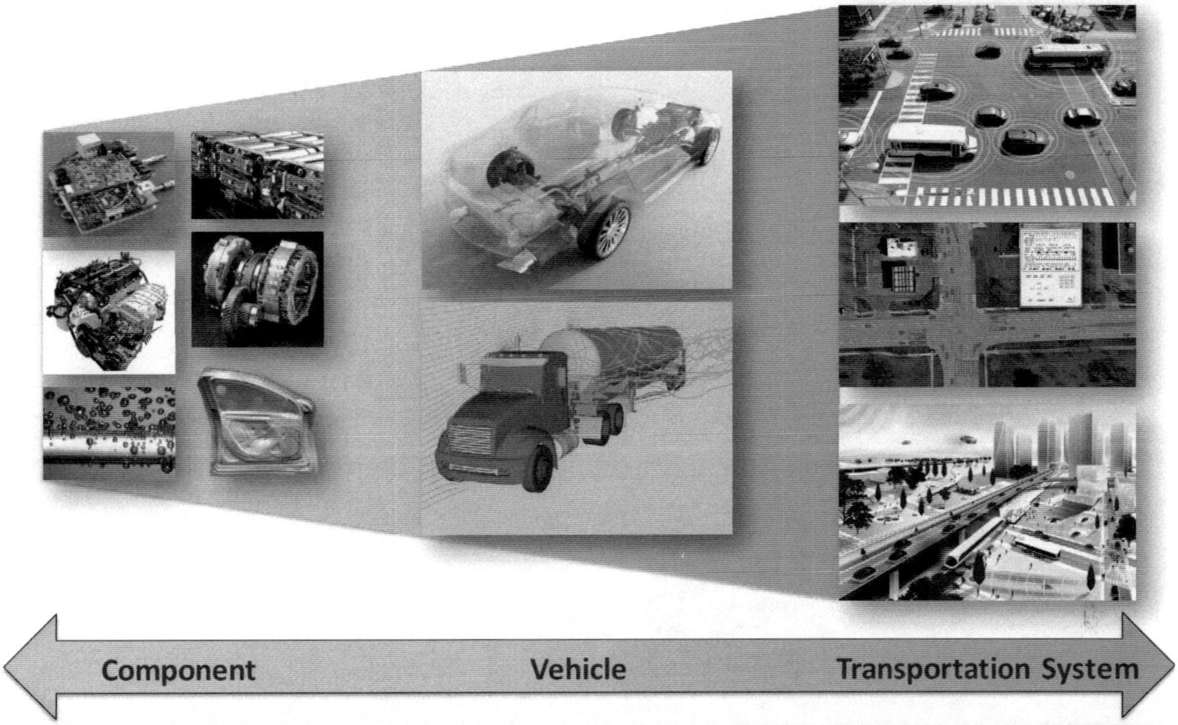

EERE'S VEHICLE TECHNOLOGIES OFFICE (VTO)

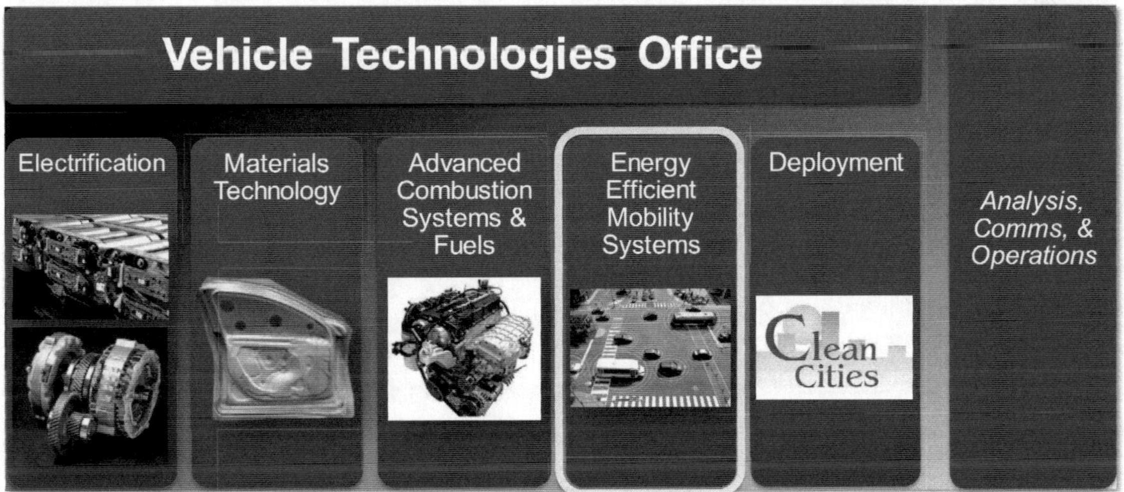

VTO develops advanced transportation technologies that:

- ✓ Improve energy ***efficiency***
- ✓ Increase domestic energy ***security***
- ✓ Reduce operating ***cost*** for consumers & business
- ✓ Improve global ***competitiveness*** of US economy

ENERGY EFFICIENT MOBILITY SYSTEMS (EEMS)

SMART Mobility Lab Consortium

High-Performance Computing / Big Data Analytics

Advanced R&D Projects

EEMS Living Labs

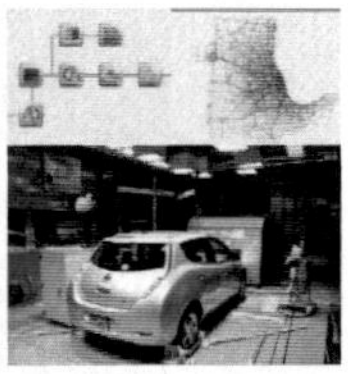

Core VTO Evaluation & Simulation Tools

SMART MOBILITY LAB CONSORTIUM

U.S. DEPARTMENT OF ENERGY
SMARTMOBILITY
Systems and Modeling for Accelerated Research in Transportation

Argonne INL BERKELEY LAB OAK RIDGE NREL

Multi-Lab Consortium creating new knowledge and understanding about the energy implications and opportunities from future mobility.

SMART MOBILITY LAB CONSORTIUM

U.S. DEPARTMENT OF ENERGY

SMARTMOBILITY

Systems and Modeling for Accelerated Research in Transportation

Argonne · INL · Berkeley Lab · Oak Ridge · NREL

Multi-Lab Consortium creating new knowledge and understanding about the energy implications and opportunities from future mobility.

- Connected & Automated Vehicles

Connected/Automated Vehicles

- What are the ***energy, technology, and usage implications*** of connected & autonomous technologies?
- How will these systems ***operate in the real world?***
- What are the critical ***levers to promote "eco-CAV" solutions?***

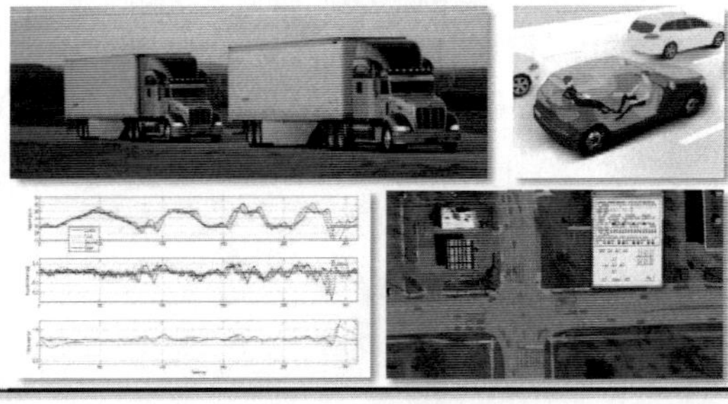

SMART MOBILITY LAB CONSORTIUM

U.S. DEPARTMENT OF ENERGY

SMARTMOBILITY

Systems and Modeling for Accelerated Research in Transportation

Argonne · INL · Berkeley Lab · Oak Ridge · NREL

Multi-Lab Consortium creating new knowledge and understanding about the energy implications and opportunities from future mobility.

- Connected & Automated Vehicles
- Mobility Decision Science

Mobility Decision Science

- What are the transportation energy ***impacts of potential lifestyle trajectories?***
- How do consumers and companies make ***travel decisions*** in the short / medium / long-term?
- What mechanisms are available to ***influence consumer decisions?***

SMART MOBILITY LAB CONSORTIUM

U.S. DEPARTMENT OF ENERGY
SMARTMOBILITY
Systems and Modeling for Accelerated Research in Transportation

Argonne · INL · Berkeley Lab · Oak Ridge · NREL

Multi-Lab Consortium creating new knowledge and understanding about the energy implications and opportunities from future mobility.

- Connected & Automated Vehicles
- Mobility Decision Science
- Urban Science

Urban Science

- ***How will SMART-enabled mobility impact the urban traveler*** in terms of VMT, congestion, vehicle ownership, mobility-as-a-service?
- What are the long-term ***impacts on the urban built environment?***
- What are the energy impacts of ***optimized signal management and automated mobility districts?***

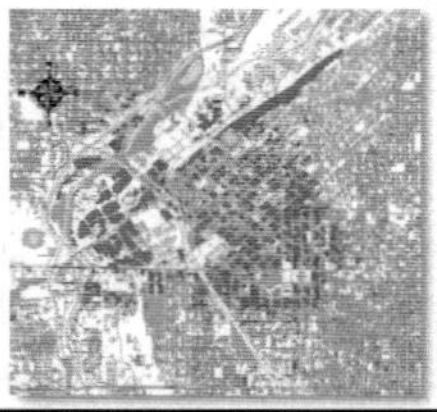

SMART MOBILITY LAB CONSORTIUM

U.S. DEPARTMENT OF ENERGY
SMARTMOBILITY
Systems and Modeling for Accelerated Research in Transportation

Argonne · INL · Berkeley Lab · Oak Ridge · NREL

Multi-Lab Consortium creating new knowledge and understanding about the energy implications and opportunities from future mobility.

- Connected & Automated Vehicles
- Mobility Decision Science
- Urban Science
- Advanced Fueling Infrastructure

Advanced Fueling Infrastructure

- ***What infrastructure is required*** to support future mobility systems?
- How can next-gen fueling/charging infrastructure ***enable energy-efficient transportation?***
- What are the ***costs and benefits***, and where should infrastructure investments be made?

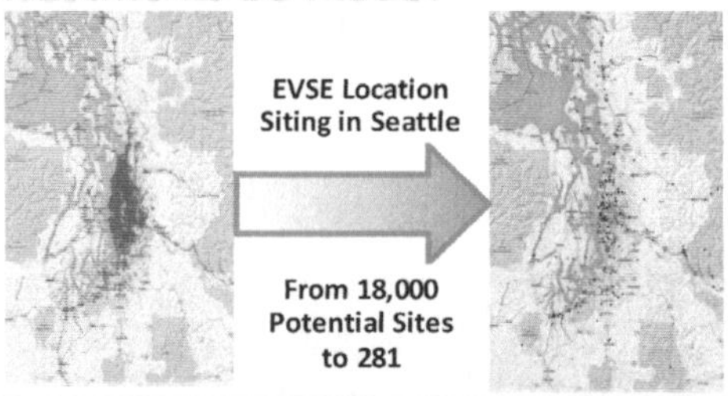

SMART MOBILITY LAB CONSORTIUM

U.S. DEPARTMENT OF ENERGY
SMARTMOBILITY
Systems and Modeling for Accelerated Research in Transportation

Argonne INL Berkeley Lab Oak Ridge NREL

Multi-Lab Consortium creating new knowledge and understanding about the energy implications and opportunities from future mobility.

- Connected & Automated Vehicles
- Mobility Decision Science
- Urban Science
- Advanced Fueling Infrastructure
- Multi-Modal Transportation

Multi-Modal Transportation

- What are the potential energy benefits of ***reduced modality interface barriers?***
- What are the interactions between ***mass transit and transportation network companies?***
- What opportunities do evolving household spending and ***commodity flow bring for freight logistics?***

SMART MOBILITY LAB CONSORTIUM

U.S. DEPARTMENT OF ENERGY
SMARTMOBILITY
Systems and Modeling for Accelerated Research in Transportation

Argonne INL Berkeley Lab Oak Ridge NREL

Multi-Lab Consortium creating new knowledge and understanding about the energy implications and opportunities from future mobility.

- Connected & Automated Vehicles
- Mobility Decision Science
- Urban Science
- Advanced Fueling Infrastructure
- Multi-Modal Transportation

Quantifying energy savings potential of vehicle connectivity and automation in merging roadway scenario (ORNL).

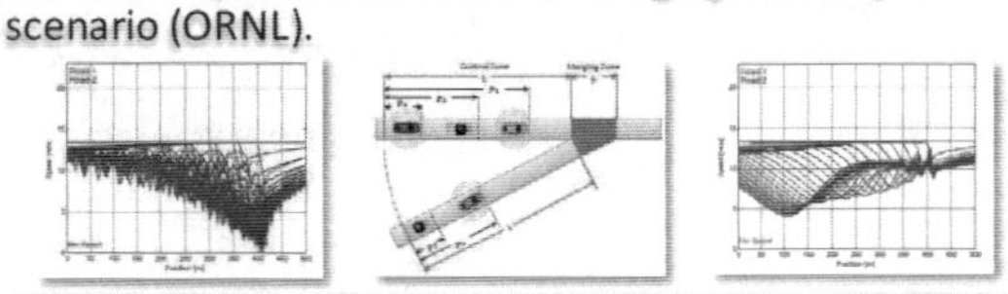

Quantifying the energy benefits of CAV-enabled drive smoothing for multiple powertrain technologies (ANL).

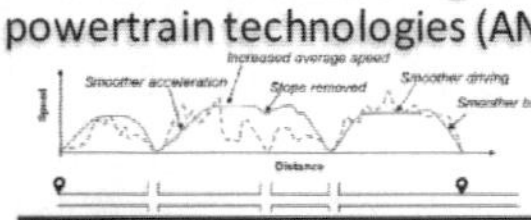

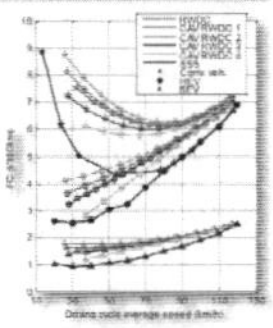

Modeling charging requirements for electrified shared mobility service fleets using spatially-resolved vehicle activity patterns (INL/NREL).

Analyzing energy impacts of autonomous driving in Chicago metropolitan area using agent-based transportation simulation (ANL).

ADVANCED R&D PROJECTS

Partner with industry and academia to research and develop mobility technology solutions that lead to energy savings.

Solutions may include:
- Hardware devices
- Software solutions
- Control systems
- Advanced sensors
- Powertrain components

ADVANCED R&D PROJECTS

Partner with industry and academia to research and develop mobility technology solutions that lead to energy savings.

Solutions may include:
- Hardware devices
- Software solutions
- Control systems
- Advanced sensors
- Powertrain components

Develop an adaptive spatio-temporal intersection control system that reduces fuel use by ~15% while improving travel time (University of Michigan).

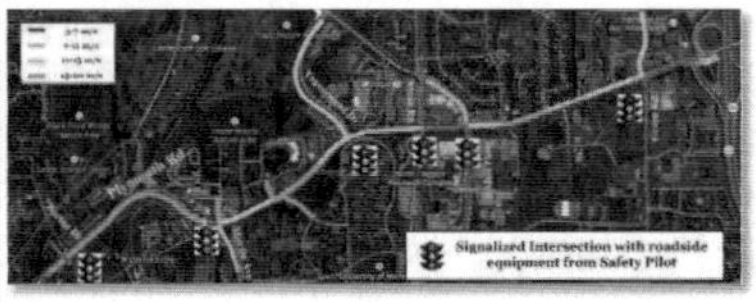

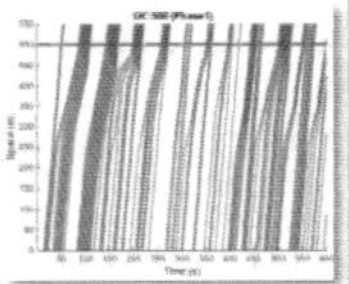

ADVANCED R&D PROJECTS

Partner with industry and academia to research and develop mobility technology solutions that lead to energy savings.

Solutions may include:

- Hardware devices
- Software solutions
- Control systems
- Advanced sensors
- Powertrain components

FOA-0001629 EEMS AWARD SELECTION

Clemson University: Boosting Energy Efficiency of Heterogeneous Connected Automated Vehicle (CAV) Fleets via Anticipative and Cooperative Vehicle Guidance

- Develop anticipative/collaborative traffic and vehicle control algorithms to achieve 10% energy savings
- High-fidelity transportation and vehicle simulation to quantify energy benefits
- On-track vehicle-in-the-loop testbed to validate results

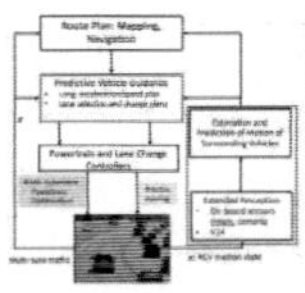

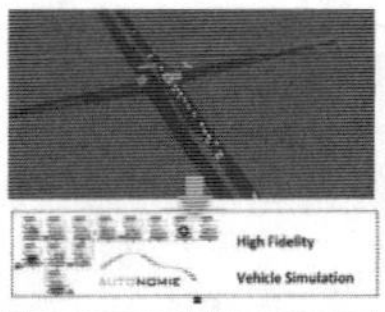

ADVANCED R&D PROJECTS

Partner with industry and academia to research and develop mobility technology solutions that lead to energy savings.

Solutions may include:

- Hardware devices
- Software solutions
- Control systems
- Advanced sensors
- Powertrain components

FOA-0001629 EEMS AWARD SELECTION

University of California – Riverside: Evaluating Energy Efficiency Opportunities from Connected and Automated Vehicles (CAVs) coupled with Shared Mobility in California

- Conduct real-world CAV data collection and analysis
- Model impacts on energy intensity (EI) and modal activity (MA)
- Construct statewide energy inventory

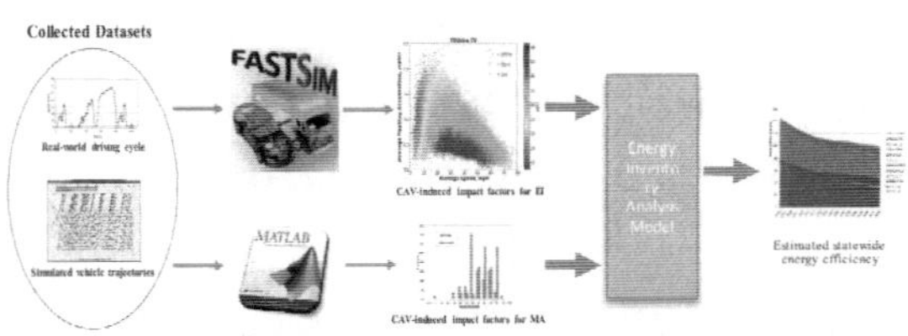

ADVANCED R&D PROJECTS

Partner with industry and academia to research and develop mobility technology solutions that lead to energy savings.

Solutions may include:

- Hardware devices
- Software solutions
- Control systems
- Advanced sensors
- Powertrain components

FOA-0001629 EEMS AWARD SELECTION

Virginia Polytechnic Institute and State University: Developing an ECO-Cooperative Automated Control System (ECO-CAC)

- Develop network-level vehicle-routing and speed-harmonization algorithms
- Develop vehicle-level energy management algorithms
- Integrate network and vehicle algorithms to demonstrate 20% energy savings while improving Level-of-Service (congestion)

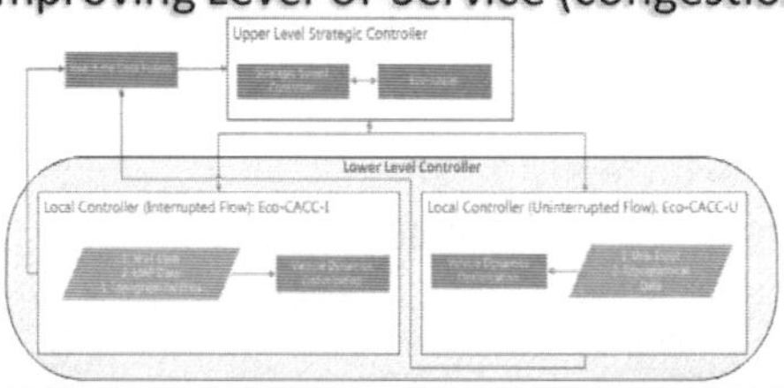

CONCLUSION

- ***Major disruption*** occurring in transportation
- ***Connected & Autonomous Vehicles*** (CAVs) are coming
- CAVs & Shared Mobility have ***dramatic implications for energy use***
- DOE must understand energy impacts and ***develop the knowledge to enable energy efficiency in transportation***

A Maximum-Mobility, Minimum-Energy Future

U.S. DEPARTMENT OF
ENERGY
Office of
ENERGY EFFICIENCY &
RENEWABLE ENERGY

David L. Anderson
U.S. Department of Energy
Vehicle Technologies Office
David.Anderson@ee.doe.gov
vehicles.energy.gov

U.S. DOT Automation Program Update

Kevin Dopart
ITS Joint Program Office
U.S. Department of Transportation

Automated Vehicles Symposium
2017

Automated Vehicle Technology Efforts Are Coordinated Across Multiple U.S. DOT Agencies

- Intelligent Transportation Systems Joint Program Office (ITS JPO)
- National Highway Traffic Safety Administration (NHTSA)
- Federal Highway Administration (FHWA)
- Federal Motor Carrier Safety Administration (FMCSA)
- Federal Transit Administration (FTA)
- Maritime Administration (MARAD)

Office of the Assistant Secretary for Research and Technology
Intelligent Transportation Systems
Joint Program Office

NHTSA

U.S. Department of Transportation
Federal Highway Administration

FMCSA
Federal Motor Carrier Safety Administration

FTA
Federal Transit Administration

Maritime Administration 1950 United States of America

NHTSA Federal Automated Vehicles Policy

- Released September 20, 2016
- Comment Period closed November 22, 2016
- Two public workshops held
- Released Policy FAQs/Clarifications, January 2017
- As noted by Secretary Chao, new guidance coming in next few months that will:
  - *Support industry innovation and encouraging open communication with the public and with stakeholders;*
  - *Make Department processes more nimble to help match the pace of private sector innovation; and*
  - *Encourage new entrants and ideas that deliver safer vehicles.*

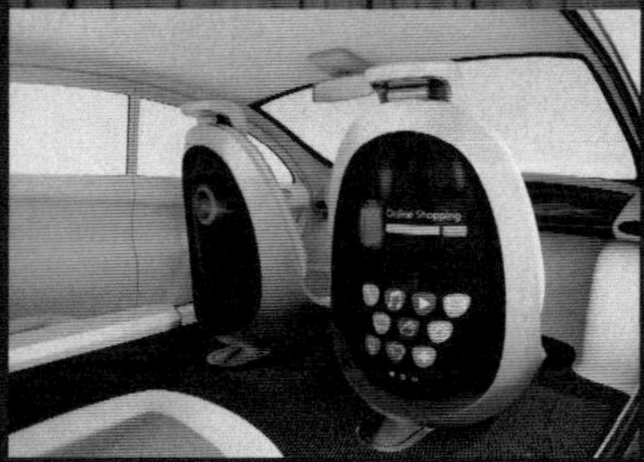

NHTSA Driving Automation Research

- Electronic Control Systems Safety
- System Performance
- Human Factors
- Cybersecurity

FHWA is Developing an Agency Wide Vision

- Automated vehicle technologies may impact the planning, design, construction, maintenance and operations of the Nation's roadway infrastructure.
- FHWA is developing an agency-wide Vision Statement to establish its role in supporting the integration of automated vehicle technologies onto the Nation's roadways.
- Vision discusses agency wide goals and guiding principles for FHWA with respect to automated vehicle technologies.

FHWA Cooperative Automation Research

Developing cooperative driving automation system applications:

Light Vehicle and Truck Platooning

- Cooperative Adaptive Cruise Control (CACC) via Vehicle-to-Vehicle (V2V)

Signalized Intersection Approach and Departure

- GlidePath via Vehicle-to-Infrastructure (V2I)

Automated Traffic Flow Optimization

- Speed Harmonization via Vehicle-to-Infrastructure (V2I)

Proof-of Concepts

- Lane Change, Merging and Weaving Operations

Evaluating impacts of automated driving technology on the transportation planning process

Developing a FTA Automation Research Plan

Overview:	5-year plan for FTA research and deployment
Lead:	FTA Office of Research, Demonstration, and Innovation
Scope:	Bus transit automation, defined broadly
Key inputs:	VAA program, MOD Sandbox, stakeholder perspectives, use case analysis, benefit-cost analysis
Target:	September 2017 (public webinar)

FTA Research Underway

Knowledge Transfer
- Inform internal and external stakeholders about state of the practice and research in progress.

Technology Transferability
- Assess transferability of light-duty and commercial vehicle sensors and technologies to transit applications.

Policy Review
- Identify and analyze federal, state, and local policies relevant to transit automation and provide recommendations for needed revisions and/or development of new policies.

Photo Credits: Volpe Center

FMCSA Automated Commercial Vehicle Activities

- Gathering information on issues relating to design, development, testing, and deployment of highly automated commercial vehicles.
- FMCSA seeks information on how to ensure that Federal safety regulations provide appropriate standards for the safe operation of HACVs from design and development through testing and deployment.
- Collecting public comments through July 17, https://www.regulations.gov/document?D=FMCSA-2017-0114-0001
- Additional Considerations:
  - Commercial driver licensing
  - Hours of service
  - Inspection, repair, and maintenance

MARAD Automation Activities

Low Speed Automated Truck Queue at Ports and Warehouses (joint project with FMSCA and JPO)

- Exploring application of automation to low-speed commercial vehicle operations at port terminals and warehouses
- Review of related studies and papers
- Surveys of industry and technology stakeholders
- Technology scan of existing or near-term enabling technologies
- Cost estimation and safety benefits

ITS JPO Automation Research

Technical Research

- Assessing Applications for Improving System Safety
- Developing Impacts Assessment
- Evaluating New Data Collection & Sharing Models

Program and Policy Support

- Strategic Planning and Roadmap Development
- Modal Policy Support and Coordination

Stakeholder Engagement

- Supporting Professional Capacity Building Efforts
- Facilitating International Coordination
- Low-speed Automated Shuttle Information Sharing Working Group

For More Information

www.its.dot.gov

Kevin Dopart
US DOT / ITS JPO
kevin.dopart@dot.gov

AUTHOR INDEX